# FREQUENCY CONVERSION

# THE WYKEHAM TECHNOLOGICAL SERIES
## for universities and institutes of technology

*General Editor:*
J. Thomson, M.A., D.Sc., F.I.E.E., F.Inst.P.
recently Chief of Technical Studies,
Sangamo Weston, Ltd.

THIS BOOK is one of a series in which authorities will attempt to introduce young people to the various technologies in the subjects discussed. Primarily these books are aimed at young graduates or apprentices who are about to begin a career in industry or who have just started work. For this reason more attention has been paid to giving a broad view of each subject than to ' necessary and sufficient ' mathematical proofs. Indeed the use of mathematics has been sparing.

Another feature of the Technological Series is the mixture of disciplines. For example, when micro-circuits are being discussed, the physical basis of semi-conduction phenomena is inextricably mixed up with the chemistry of almost pure single crystal formation and with the engineering necessary to produce a monolithic circuit. This admixture of disciplines is a feature of industrial research and development. Indeed, it might be argued that this feature of industrial work makes it very unsuitable for the so-called ' specialist ' who only recognises his own discipline.

It is hoped both by the authors and by the publishers that this series will open up some new horizons to the young scientist and engineer in industry.

# FREQUENCY CONVERSION

**Part I** – Common Conversion Devices    J. Thomson

**Part II** – Light Conversion Devices    W. E. Turk

**Part III** – Masers and Lasers    M. J. Beesley

DISTRIBUTED BY

**Springer-Verlag New York Inc.**
175 Fifth Avenue, New York, N. Y. 10010

**WYKEHAM PUBLICATIONS (LONDON) LTD**
(A subsidiary of Taylor & Francis Ltd)
**LONDON & WINCHESTER**
**1969**

*First published* 1969 *by Wykeham Publications (London) Ltd.*

*Cover illustration—A beryllia-tube argon laser (S.E.R.L. photograph).*

*Printed in Great Britain by Taylor & Francis Ltd.*
*10–14 Macklin Street, London, W.C.2*

85109 030 3

*Worldwide distribution excluding the Western Hemisphere, India, Pakistan, and Japan by Associated Book Publishers Ltd., London and Andover.*

## SI Units

This book, as far as is reasonable, makes use of SI units. For readers who are not acquainted with the system the following notes may be valuable.

The six basic SI units are:

| Name of physical quantity | Name of unit | Symbol of unit |
|---|---|---|
| length | metre | m |
| mass | kilogramme | kg |
| time | second | s |
| electric current | ampere | A |
| thermodynamic temperature | kelvin* | K |
| luminous intensity | candela | cd |

The derived units are:

| Name of physical quantity | Name of unit | Symbol of unit |
|---|---|---|
| force | newton | N |
| energy, heat | joule | J |
| power | watt | W |
| electrical charge | coulomb | C |
| electrical potential difference | volt | V |
| electrical resistance | ohm | $\Omega$ |
| electrical capacitance | farad | F |
| magnetic flux | weber | Wb |
| magnetic flux density | tesla | T |
| inductance | henry | H |
| luminous flux | lumen | lm |
| illumination | lux | lx |
| frequency | hertz | Hz |
| customary temperature | degree Celsius | °C |

and agreed prefixes and symbols used to indicate multiples of units in powers of ten are:

| Multiple | Prefix | Symbol |
|---|---|---|
| $10^9$ | giga | G |
| $10^6$ | mega | M |
| $10^3$ | kilo | k |
| $10^{-3}$ | milli | m |
| $10^{-6}$ | micro | $\mu$ |
| $10^{-9}$ | nano | n |
| $10^{-12}$ | pico | p |

*The unit kelvin (symbol K) is recommended in place of degree Kelvin (symbol °K and deg).

# CONTENTS

*Frequency conversion*

WHY should we wish to write about ' frequency conversion ' when the subject covers work in many disciplines and in apparently unrelated technologies? The answer, or at least part of it, is to be found in the word ' apparently '. As science progresses and as Man puts its discoveries to work, there becomes evident a unifying of all scientific disciplines. If, for example, a mathematician can be called a scientist (and there are some who doubt it), his work can be performed in connection with physics, chemistry, biology and all kinds of engineering, economics and accountancy. This kind of universality of mathematics springs from the fact that in all sciences logical thinking is required.

In the last fifty years physicists have been chasing mathematicians throughout nearly all the sciences and some of them have claimed that all methodical thinking is physics. While we cannot go along with that, we believe it is true that the physicists have acted as catalysts bringing together the different disciplines of sciences and at least partly fusing them together.

So it is with our immediate subject. As we shall show, there is a real bond between all the diverse examples of frequency conversion we shall discuss. Let us examine some of these in a preliminary fashion so that we know what to look for in succeeding parts of this book.

*Frequency conversion in music*

This is probably the oldest example we have of frequency conversion. When a stringed instrument is played, either by plucking, as with the guitar, or by bowing as with the violin, or by means of hammers as with the pianoforte, the peculiar timbre of the instrument depends very largely on the harmonics of the fundamental note which are developed. The beauty of tone of a great violin has been achieved by shaping the instrument in such a way that all the harmonics observable by the human ear are produced in the correct proportions. The range of the ear is enormous—from about 50 Hz to almost 20 kHz—and the sub-harmonics and harmonics of a Stradivarius violin can cover the whole of this range. The pianoforte, although not as perfect an instrument as the violin (because it has only a limited number of keys) surpasses it in one respect. When it is properly constructed the ' sympathetic ' tones, brought out from its individual strings, can give a greater fullness of tone than the violin can produce. Again, the positions of the hammers are adjusted to give the proper proportion of harmonics.

In all musical instruments we obtain harmonics, unless perhaps in the concert organ when the latter is played softly. These harmonics are produced by forcing the strings or the pipes to resonate strongly—one might almost say too strongly. When a piano is played softly the resultant note is purer and so with all the other instruments mentioned. The harmonics are the effect of forcing the resonators, reeds, strings, etc., beyond the point where their response is 'linear'. What do we mean by 'linear response'? The best way is to illustrate it by a diagram (fig. I.1), where the output of a violin is graphed against the input.

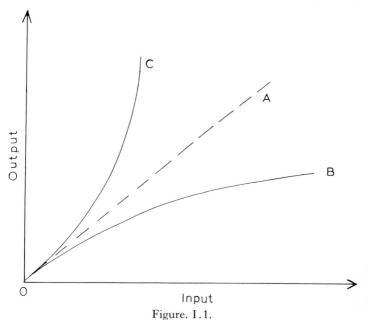

Figure. I.1.

If the output is proportional to the input the response is said to be linear. This is shown by the line OA. If the response falls below OA (as in line OB) or rises above OA (as in line OC) the response is non-linear. It is this non-linear response which produces the harmonics and the characteristic tone of a first-class violin.

But there is more to it than that! If the violin is shaped properly— if the belly and the back are of the correct thickness and are made of the right wood—and if the sounding post is in the correct position, the violin filters the harmonics so that the resultant noise is the most pleasing to the ear. So the violin not only has a non-linear response, it also acts as a composite sound filter. Is it any wonder that it is difficult to copy a Cremona violin?

*Frequency conversion in radio communications* (Part I of this book)

Most of us have radio receivers, using ' radio ' in its popular sense as opposed to ' television '. These radio receivers are almost invariably what is known as ' superheterodynes ', i.e. they depend on a conversion from the frequency of the incoming signal to a fixed frequency (usually about 450 kHz) at which most of the amplification is done.

If we examine the operation of such a receiver we shall find that the frequency conversion is achieved by utilizing the ' non-linear ' characteristic of a transistor (or vacuum tube). So here again in an entirely different field from music we find that frequency conversion requires a non-linear action.

The radio ' ham ', or the amateur radio enthusiast, transmits as well as receives. The frequency of his transmitter is generally controlled by a quartz crystal, but the frequency of the latter may be eight, nine or twelve times smaller than his transmitting frequency. To obtain such ' frequency multiplication '—a special kind of frequency conversion— use is again made of the non-linear characteristics of vacuum tubes or transistors.

*Frequency conversion in television* (Part II of this book)

The signal picked up by a television receiver is at a very much higher frequency than that normally used for ' radio ' broadcasting. This is for two inter-connected reasons. First, there is no space left at the lower frequencies to accommodate television signals and, secondly, connected with that, television broadcast signals require a very much greater frequency band than do radio broadcast signals. For example, BBC 2 requires about 12 MHz at a frequency of over 600 MHz as compared with the medium wave radio broadcast bands which are only a few kHz wide at frequencies of the order of 1 MHz. These figures are only examples.

When the signals are received they are in two separate channels—the vision channel and the sound channel. So far as the sound channel is concerned, the receiver is a conventional ' super-het '. But the vision signal has to operate many different devices in the receiver. First, it has to trigger and maintain the raster of lines across the television tube and here is frequency conversion with a vengeance. How this is done is explained in a later section. At the moment all we can say is that it is a very non-linear procedure.

Secondly, the intensity of the spot which is flying across the tube, giving 405 or 625 lines on the face of the tube, is varied in accordance with the picture—a bright patch in the scene giving a bright patch on the tube and a dark patch giving a dark one. Again, this is a frequency conversion—the energy of the light wave is varying according to a voltage, or, if you like, the relatively low frequency voltage variation is producing a variation at a frequency billions of times higher. This is the most extreme frequency conversion—that which occurs in the television tube when the h.t. voltage causes electrons to strike the screen and to produce phosphorescence of the material on its inside surface. The phosphorescence itself is another example of frequency conversion.

At the television transmitter the opposite effect is used. The light entering the camera is transformed in the camera to a voltage which (in black and white TV) modulates the radio wave which is transmitted. The transformation is brought about by photoelectric action.

*Frequency conversion in laser action* (Part III of this book)

Lasers are one of the latest of Man's discoveries in science and it would be unwise to try to describe how a laser works without a reasonable amount of preliminary explanation. Suffice it to say that a laser beam is a beam (emanating from a solid, liquid or gas) which may be in the infra-red region of the spectrum or in the visible region or in the ultra-violet region. Its most important characteristic is its narrow band-width indicating that the radiation is extremely *coherent*. By the term coherent we mean that a laser beam is a very good approximation to the theoretical description of electromagnetic radiation which pictures the wave as having a definite frequency and suffering no sudden changes of phase over the whole of its path—no matter how long that path may be. In addition, the wave-front of the beam must show only a small or regular change in phase across its surface.

The term laser is derived from *L*ight *A*mplification by *S*timulated *E*mission of *R*adiation and the frequency conversion occurs as follows. First, the laser is excited by an external source of energy which may be at any one of a variety of different frequencies ranging from ultra-violet light through radio frequencies down to a d.c. discharge. Whatever the external source, the result is the same: atoms in the laser material are excited and in falling back to their ground state undergo a transition between two energy levels where conditions for amplification are satisfied. Energy then builds up at a frequency depending on the particular energy gap. The addition of mirrors to each end of the laser material increases the amplification and by making one mirror semi-transparent, a continuous beam of radiation escapes. This is the coherent laser beam.

Clearly every laser shows frequency conversion. This is, at the moment, the most sophisticated application of our basic principle.

Having introduced the reader to examples of the use of the principles of frequency conversion in a variety of technologies, these subjects will be examined in more detail in what follows.

# FREQUENCY CONVERSION

## Common Conversion Devices Part I

J. Thomson
Editor, International Journal of Electronics

## frequency conversion in radio transmitters and receivers

THE conventional block diagram of a single channel radio transmitter is exemplified by fig. 1.1.

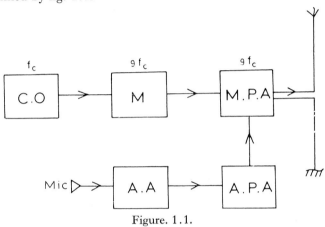

Figure. 1.1.

A crystal-controlled oscillator is generally the source of the radio carrier wave. For example, it may give a sinusoidal output at a frequency of 1 MHz (1 megacycles per second, or 1 000 000 cycles per second). This we have labelled $f_c$ in the figure.

The crystal oscillator frequency is then often multiplied by a factor of 8 or 9 or 12 to obtain the final carrier frequency. In the figure we have assumed that the multiplying factor is 9 and that the carrier frequency is 9 MHz. In fact the multiplying factor may be anything we please, but it is usually $2N \times 3M$, where $N$ may be $\frac{1}{2}$ or an integer and $M$ may be $\frac{1}{3}$ or an integer. In fig. 1.1, $N = \frac{1}{2}$ and $M = 3$.

The carrier is then amplified to the required level, using as many stages of Class C amplification as are necessary. The last stage (and sometimes the second last) are then modulated by the audio-frequency signal. In the figure the multiplying stages are represented by M and the modulated power amplifier stages by M.P.A. The audio power amplifier (A.P.A.) which feeds the M.P.A. receives its input power from audio amplifiers (A.A.) which are themselves fed from a microphone.

Note that in this book the figures are numbered separately for each part.

# 1. *Frequency multiplication*

This is a special case of frequency changing. It is achieved by using the non-linear part of a vacuum tube or transistor characteristic. In either case the active component is biased until only the positive (or negative) peaks of the oscillation cause the tube or transistor to carry current. Figure 1.2 (*a*) shows a typical characteristic for a vacuum tube and fig. 1.2 (*b*) shows the effect its input signal has at its output.

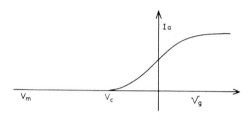

Figure 1. 2*a*.

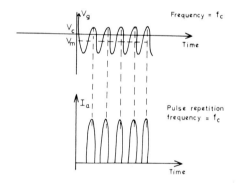

Figure 1.2 *b*.

The negative voltage at which the anode current becomes zero is $V_c$. For multiplication by 3 the tube is biased to $V_m$, where $V_m$ is approximately $3V_c$. Turning now to (*b*), the input signal is shown as a sinusoid in which only the parts above $V_c$ are effective in producing a current in the anode circuit. The spikes of anode current are also shown. These occur with a pulse repetition frequency of $f_c$. The multiplying circuit is similar to that of a Class C power amplifier except in the tuning of the anode parallel resonant circuit. It is shown in fig. 1.3.

The circuit $LC$ is tuned to frequency $3f_c$, so that at every third oscillation it receives an impetus from the anode current. Consequently the relation between the input and output waves is as shown in fig. 1.4.

It will be noted that there is a certain damping of the output wave at $3f_c$. This damping can be very considerably reduced by using circuits which have a high $Q$, when the damping is difficult to detect.

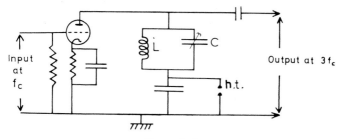

Figure 1.3.

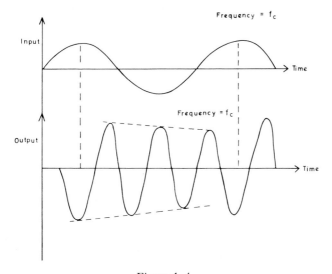

Figure 1.4.

## 2. *High level amplitude modulation*

This is another example of frequency changing, although in this case the change is apparently brought about without the use of a non-linear characteristic. The basic circuit for producing high level amplitude modulation is shown in fig. 1.5. In practice this circuit is modified in many respects to avoid distortion.

The triode circuit is a Class C amplifier, but the anode supply is fed through the secondary of an audio-frequency transformer which provides power. The capacitor C must be large enough to by-pass the radio frequency, but not large enough seriously to affect the audio frequency.

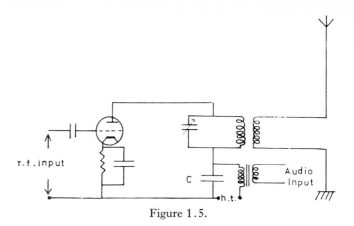

Figure 1.5.

Then, if the audio power is large enough, the modulated wave emitted by the aerial is of the form:

$$E = \sin 2\pi ft + \sin 2\pi f_m t \sin 2\pi ft, \qquad (1.1)$$

when an audio wave of frequency $f_m$ is fed into the transformer. The radio frequency is $f$ and the radio wave is 100% modulated by the audio, i.e. its amplitude varies from zero to 2 in the case quoted in equation (1.1).

Transforming the expression for $E$, and remembering that $2 \sin \theta \sin \phi = \cos (\theta - \phi) - \cos (\theta + \phi)$:

$$E = \sin 2\pi ft + \tfrac{1}{2} \cos 2\pi(f - f_m)t - \tfrac{1}{2} \cos 2\pi(f + f_m)t. \qquad (1.2)$$

This shows that $E$ is made up of three sinusoidal waves, one at frequency $f$, another at frequency $(f - f_m)$ and a third at frequency $(f + f_m)$.

If the highest frequency to be transmitted is 3 kHz—as would be the case with a single speech channel—the band-width of the transmitter would need to be at least 6 kHz, centred on frequency $f$, always assuming that the transmitter is crystal controlled, so that the variations in $f$ are negligible.

As a matter of convenience the expression for $E$ given in equation (1.1) may be classified as non-linear, since it is of the form $(1 + x)y$ and both $x$ and $y$ vary. This suggests that a linear equation may be defined quite rigorously as one having the form:

$$Z_e = A + Bx + Cy + \ldots, \qquad (1.3)$$

where $A$ and $B$ are constants and $x$ and $y$ are the variables.

Any other equation, such as

$$Z_{n1} = A + Bx + Cx^2 + \ldots \qquad (1.4)$$

or

$$Z_{n2} = A + Bx + Cy + Dxy + Exy^2 + \ldots, \qquad (1.5)$$

expresses a non-linear relationship.

10

In equations (1.3), (1.4) and (1.5) only two variables have been used. There may, of course, be more than two, but in almost all cases which will be cited, two independent parameters will suffice.

### 3. *Superheterodyne receivers*

As mentioned in the introduction, most radio receivers nowadays are 'superhets'. The block diagram for a typical receiver of this kind is shown in fig. 1.6.

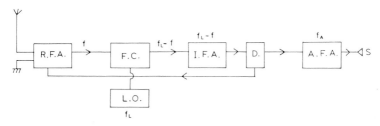

Figure 1.6.

Starting at the left of the diagram the signal is received by an aerial and (in good class receivers) undergoes a small amount of amplification at its original frequency, $f$. It then passes into a frequency changer (F.C.) where it is mixed with the frequency of a local oscillator (L.O.). The resulting signal is at the frequency $(f_L - f)$ and this is constant over the tuning band of the receiver, because the L.O. is 'ganged' to the R.F.A. so that an increase in $f$ of, let us say, 10 kHz, gives rise in an increase in $f_L$ of 10 kHz. Because $f_L - f$ is fixed (it is usually about 450 kHz for broadcast receivers) the amplification can be achieved in the most economical fashion possible and in the intermediate frequency amplifier (I.F.A.) it is not unusual to find a power gain of $10^8$.

Incidentally, radio engineers express power gains in decibels. This is the measure of the ratio of the output power, $P_0$, to the input power, $P_I$. The equation is:

$$\text{gain in decibels} = 10 \log_{10} \left( \frac{P_0}{P_I} \right), \tag{1.6}$$

showing that a gain of $10^8$ is 80 decibels or 80 dB.

This relatively large signal then goes to the detector, D, where the modulation is taken out and amplified in the audio-frequency amplifier (A.F.A.). There is one feedback loop from the detector to the radio-frequency amplifier. This reduces the amplification when the signal is strong and increases the amplification when it is weak, so removing much of the effect of fading. The feedback loop gives automatic gain control (A.G.C.). Finally, the audio signal is fed to a loud-speaker, S.

11

## 4. *The frequency changer*

It is now necessary to take a closer look at the frequency changer. This can take one of two forms. In the first case the signal from the preceding amplifier and the signal from the local oscillator may be applied to two electrodes of a vacuum tube or transistor. A typical transistor circuit is shown in fig. 1.7.

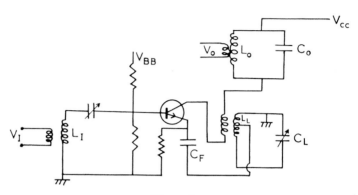

Figure 1.7.

The input signal is $V_I$ and it is applied to the base of the transistor. The remainder of the circuit makes up the local oscillator by reason of the tuned circuit $L_L C_L$ and the feedback through the capacitance, $C_F$. Consequently the collector current depends both on the local oscillation and on the input signal and to a first approximation is given by the equation:

$$I_c = A + B \sin 2\pi f t + C \sin 2\pi f_L t + D \sin 2\pi f t \sin 2\pi f_L t.$$

The last term on the right-hand side may be expanded (as we have seen in equation (1.2)) to give:

$$\tfrac{1}{2} D \cos (f_L - f) t - \cos (f_L + f) t \ldots \text{(part of (1.5))}$$

and so the principal components of the current in the collector circuit are at frequencies $f$, $f_L$, $f_L - f$ and $f_L + f$. The parallel tuned circuit $L_0 C_0$ resonates at the frequency $(f_L - f)$ and so all other frequencies are greatly diminished. Indeed, the higher frequencies $f_L$ and $f_0$ are by-passed via capacitance $C_0$. The output voltage, $V_0$, is therefore at the intermediate frequency which, for broadcasting, is about 450 kHz.

## 5. *The detector*

After the signal at frequency $(f_L - f)$ has passed through the I.F. amplifier it is demodulated by the detector. The action is illustrated by fig. 1.8.

12

By using the non-linear part of a transistor or vacuum tube device, the positive peaks of the modulated wave are accentuated while the negative peaks are diminished (or vice versa) and this distortion brings to light the modulation at frequencies which are low compared with the

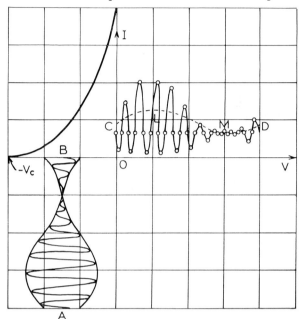

Figure 1.8.*

I.F. frequency. These low frequencies are fed into a transformer and the I.F. signals are by-passed to earth in the usual fashion. This is an example of the application of equation (1.4) which represents the $I$, $V$ characteristic of the device depicted in fig. 1.8.

If $I$ is the collector current in a transistor the base voltage of which, $V_B$, varies with the I.F. voltage, then

$$I = A + BV_B + CV_B^2 + \ldots.$$

Considering only the expression $CV_B^2$,

$$CV_B^2 = C\{-a + V_{I.F.} \sin 2\pi(f_L - f)t[1 + m \sin 2\pi f_m t]\}^2. \quad (1.7)$$

The second term inside the curly brackets gives rise to:
$$CV_{I.F.}^2 \sin^2 2\pi(f_L - f)t[1 + m \sin 2\pi f_m t]^2$$

$$= \frac{C}{2} V_{I.F.}^2[1 - \cos 4\pi(f_L - f)t][1 + 2m \sin 2\pi f_m t + m^2 \sin^2 2\pi f_m t].$$

Multiplying these two factors, gives as one of the terms

$$CV_{I.F.}^2 m \sin 2\pi f_m t,$$

* Fig. 7.5, p. 108, *Services Text-Book of Radio*, Volume III, by permission of H.M. Stationery Office.

showing that the modulation appears as a new frequency after the distortion produced by the device. (The parameter $m$ is used in these equations to indicate the depth of modulation. It was omitted from the description of high level modulation for the sake of simplicity.)

Unfortunately, the distortion does not stop here. Another term in the expression is:

$$\tfrac{1}{2}CV_{\text{I.F.}}{}^2 m^2 \sin^2 2\pi f_{\text{m}}t,$$

which is

$$\tfrac{1}{4}CV_{\text{I.F.}}{}^2 m^2[1 - \cos 4\pi f_{\text{m}}t],$$

one element of which is $(CV_{\text{I.F.}}{}^2 m^2/4)(\cos 4\pi f_{\text{m}}t)$ or second harmonic distortion. There is also third, fourth, etc., harmonic distortion depending on the exact shape of the characteristic.

Some of this distortion can be eliminated by using what is wrongfully called a linear demodulator. The latter (a crystal diode is a good example) has a characteristic which closely resembles that of fig. 1.9.

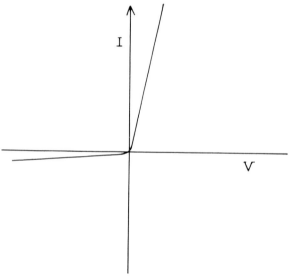

Figure 1.9.

Clearly, if the I.F. modulated wave is applied as the voltage to this, a modulation will appear and the second harmonic distortion will be greatly reduced. The characteristic is, however, a non-linear one (speaking mathematically) because of the limited length of the two straight lines.

It will be seen then that there are two frequency changes in the modern superheterodyne receiver, although one of these is usually known as detection. The action is the same in both.

14

# CHAPTER 1.2

## frequency conversion in other branches of electronics

ALTHOUGH it is not intended that this chapter should contain all applications of electronic frequency changing, some of the more interesting will be considered, excluding television and lasers.

### 1. *Horology*

One of the more common examples is taken from modern horology Until comparatively recently clocks were beautiful examples of precision workmanship embodying many of the principles of mechanical engineering. All such clocks were mechanical; occasionally one came across a master clock which gave electrical impulses to a number of 'slave' clocks, but the mechanism here was simple, the chief problem being to make certain that the pendulum's energy (kinetic + potential) was of such a magnitude that the small amount bled off to control the slaves could either be neglected or could be accounted for by decreasing the length of the pendulum.

The coming of the crystal controlled oscillator was the first serious attack on the mechanical clock. Once technologists had realized that the frequency of a 'crystal oscillator' could be kept within 1 or 2 parts in $10^8$ of its nominal frequency, the mechanical clock as a sub-standard for time measurement was doomed. The discovery that certain lines in the caesium spectrum could be used to establish an absolute standard of frequency and therefore of time gave the death-blow to the mechanical clock and dealt a serious blow to the crystal oscillator.

A few figures may be quoted. As measured by the caesium clock, the length of the day varies in a fairly regular fashion by about 1 millisecond (ms) in the course of a year, being shortest towards the middle of the year (July) and longest in March or April and in November, with a minor dip between November and March. In addition to this roughly periodic variation there is a smaller drift towards a longer day, amounting to about 1 ms in nine years. The mean solar day contains 86 400 mean solar seconds and so the measurement of a change of 1 ms in three months represents an accuracy of observation of 1 part in $10^3 \times 86\,400 \times 91$ or 1 in $10^{10}$. The frequency of the caesium clock is 9 192 631 770 Hz $\pm\,0\cdot1$ Hz, which means that it is known to 1 part in $10^{10}$, agreeing with the previous calculation. This is 1000 times more accurate than could be achieved using astronomical time, since the best of mechanical clocks, after all due allowances had been made for their

15

rate, were not any more accurate than the astronomical time on which they were based!

The quartz clock depends for its accuracy on the stability of a piece of quartz which, needless to say, is maintained at a very constant temperature and is not subjected to any appreciable strain. The best long-term accuracy which can be achieved using the quartz oscillator is about 1 part in $10^8$, still a long way behind the caesium clock. Moreover, there is one important difference between them. While the quartz oscillator must be calibrated, either by the use of astronomical time or the caesium clock, the latter is its own standard. The frequency of the hyperfine line of caesium can be calculated and checked by observations on other lines in the same spectrum. So we have now two standards of time—the mean sidereal day and the frequency of the line from caesium vapour.

This is an awkward situation, where we have two inter-related standards. A compromise solution which appears to suit everyone is that the frequency of the caesium clock is taken to be as nearly as possible equal to its value in terms of the average value of the mean solar second over the period during which reliable data have been recorded. Time, calculated on this basis, is known as Ephemeris Time and the resulting frequency of the caesium line is 9 192 631 770 $\pm$ 20 Hz. It should be noted that the uncertainty of measurement of this frequency is less than $\pm 0.1$ Hz. The $\pm 20$ Hz is necessary in view of the un-certainties inherent in Ephemeris Time measurement.

This matter deserves further consideration. The concept of time was given the status of a physical dimension which could be measured by the fact that the Earth appeared to rotate at a uniform speed on its axis. One complete rotation was known as the sidereal day, since it could be defined as the interval between two successive transits across the meridian of any star. This unit could not be used in everyday life because it was not related to sunrise and sunset and so the mean solar day was taken as the practical standard. In terms of the mean solar day, the sidereal day is approximately 23 hours, 56 min, 4.05 sec.

The practical standard (the mean solar day) was divided into $24 \times 60 \times 60$ sec or 86 400 sec and the mean solar second was the basic standard of time for many years. The velocity of light was measured in terms of this unit and a great number of physical constants, both old and new, depended upon it.

As we have seen, it is only recently that the mean solar second has been found to vary, both periodically and aperiodically.

At first sight it appears to be impossible to set up any other standard. For example, the frequencies of all spectral lines can only be measured, using the velocity of light *in vacuo* as one of the terms in the equation, or alternatively, using Planck's constant of action, $h$, which again involves the measurement of time ($h$ is in joule sec), so, each frequency is based on a variable unit of time.

What has happened is that the *new* standard of time has been *defined* as 9 192 631 770 Hz with an uncertainty of $\pm 0.1$ Hz, being the frequency of the caesium line, and all other constants *could* be based on this unit. However, it is wholly unnecessary to recalculate existing constants since none of them are appreciably affected by the change. Ephemeris time which is necessary for practical reasons has to accept an uncertainty in terms of the fundamental unit of $\pm 20$ Hz.

Whether we start with the caesium clock or with the quartz clock, the fundamental interval of time which they measure is much too short to be of practical use in measuring time intervals of the order of 1 sec or longer. The simple way out of this difficulty is to divide the frequency by a series of integers, until we arrive at a satisfactory unit. For example, suppose our quartz oscillator is at a frequency of 100 kHz (which is fairly common). Then it is easy to divide this frequency by 10 five times, thus arriving at a single pulse every second. This can be achieved by several methods, two of which will now be described briefly.

## 2. *Frequency division*

This can be carried out using a trochotron, a dekatron or a ring counter and of these the modern method is to use a ring counter. Since the latter is generally of the binary type, the action of a typical dekatron circuit will first be described. (Note: The name dekatron is the trade name given by Ericsson, Ltd. to their cold-cathode gas-filled counters. We use it merely for convenience as a shortened form of the rather elaborate scientific name. But all that is said about the dekatron is equally applicable to other cold-cathode gas-filled counters.)

The dekatron is a cold-cathode gas tube, containing argon or some other inert gas at a low pressure. The cylindrical anode is in the middle of the tube as shown in fig. 2.1. The anode is surrounded by thirty rod-shaped electrodes, which are given different names in different countries. In the U.S.A. they are known as digit cathodes (10), first intermediate cathodes (10) and second intermediate cathodes (10). All the first intermediate cathodes are connected together inside the tube as are all the second intermediate cathodes, while all except one of the digit cathodes are similarly connected together.

In any glow discharge tube the potential required to initiate the glow is greater than the potential required to maintain the glow and this fact accounts for the mode of operation of the dekatron. When the anode potential, $V_A$, is applied to the tube via a resistor, a glow is initiated on one of the digit cathodes. The potential between all the digit cathodes and the anode falls to $V_M$, where $V_M$ is the maintenance voltage of the discharge. Consequently the glow remains on the original cathode. If a negative signal is now applied to all first intermediate cathodes, the increased potential difference between the anode and the intermediate cathodes causes the discharge to move to the first intermediate next to the digit cathode which originally carried the glow. The discharge moves to this particular electrode because the ionization of the gas which occurs during a discharge is most intense in the neighbourhood of the intermediate cathode adjacent to the digit cathode originally carrying the glow. If a negative signal is now applied to the second intermediate cathodes, the glow shifts one cathode further and when the negative signal is removed the glow returns to the second digit cathode. This represents a count of 1. The circuitry associated with the dekatron comprises a phase shift network arranged so that the negative pulse is first applied to the first intermediate cathodes and then to the second.

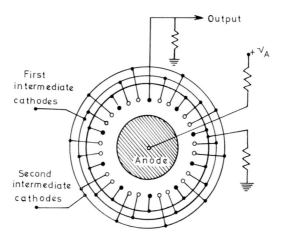

*From Richard's "Electronic Digital Components Circuits", Copyright 1967, D. Van Nostrand Company, Inc., Princeton, N.J.*

Figure 2.1.

When the pulse ceases the glow has moved by one digit. Each successive negative pulse moves the glow round in a clock-wise direction until at the tenth pulse it arrives back at the 0 digit cathode. This is independently connected to earth through a resistance, the voltage across which constitutes the output of the dekatron.

By using a bank of dekatrons connected in sequence as shown in fig. 2.2 we can count tens, hundreds and thousands of pulses. Usually a pulse-shaping circuit is included between successive dekatrons to ensure a rapid rise time of the signal.

Clearly, such a bank of counters can be made to count the time which elapses between 10, 100, 1000 or, in fact $10^n$ pulses, so, if the original pulses came from a 10 kHz oscillator, by using dekatrons we could obtain a clock which told the time in 1 sec intervals. There is one difficulty

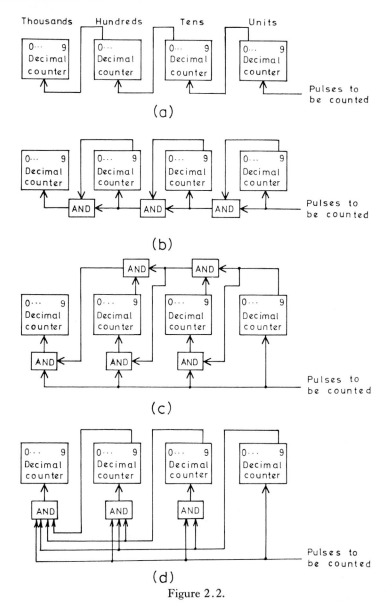

Figure 2.2.

which should be mentioned. Dekatron tubes have an upper limit of pulse repetition frequency. The best that have so far been made will not operate beyond 10 kHz.

The more modern way to obtain frequency division is by means of transistor circuits. The matter is too involved to allow of detailed discussion here, but all decimal counters operate in much the same manner as the multi-dekatron circuit described earlier. In effect, such decimal counters comprise a series of ring counters and a series of shift registers which allow a count of 10 in any part of the system to be transferred to the next higher part as 1. The big advantage of using transistors to perform the arithmetic is that the speed of counting can be increased by a factor of 10 000, allowing us to count pulses occurring 10 nanoseconds (1 ns $= 10^{-9}$ s) apart.

## 3. *The two standards of time*

One nanosecond represents a large number of pulses at the frequency of the caesium clock, i.e. at approximately $9 \times 10^9$ Hz or 9 gigacycles/sec (9 GHz). There are various methods which can be used to obtain an accurate comparison between the astronomical clock and the caesium clock. One method which theoretically gives a very high degree of accuracy is to start with a quartz crystal bar which will operate in an appropriate circuit at a frequency of about 100 kHz. This is then used as an intermediate stepping-stone. By means of frequency multiplication the frequency of the quartz crystal is stepped up to about 9 GHz, the last few stages giving each a multiplication by 3 or 5, so that there can be no doubt about which harmonic is being used.

The two signals—one obtained from the quartz oscillator and the other from the caesium clock—are then mixed in an appropriate wave-guide circuit incorporating a diode, the non-linear characteristic of which gives a signal at the difference frequency. This difference frequency can be accurately measured, since it is at a comparatively low frequency.

By this means the astronomical clock may be compared (*a*) with the crystal oscillator and (*b*) with the caesium clock. It is this type of comparison which leads to the result already quoted, that the frequency of the caesium line is 9 192 631 770 Hz $\pm 0 \cdot 1$ Hz. The crystal oscillator has a short-term stability better than 2 parts in $10^{11}$, although its long-term stability may not be better than 1 part in $10^8$.

This description ignores many of the difficulties inherent in comparing two 'standards'. For example, the astronomical clock ' second ' is one second of ephemeris time, which, as has been mentioned, is an average value over a large number of years.

There are obviously modifications to this method. Using fast counters, frequency division can be accomplished at frequencies of the order of 100 MHz, and using ' modal crystals ', i.e. crystals which will oscillate on a mechanical harmonic of their fundamental frequency, a crystal-controlled oscillation can be obtained at a frequency of 70–100 MHz and harmonics up to 9 GHz can be obtained. Only the scale of operation is different; the principle is the same. So we see that frequency conversion is an essential feature of modern horology.

## 4. *Transistor logic*

The development of electronic digital computers has to a large extent been conditioned by developments in transistors and, as computers grew in size, particularly as regards storage and arithmetic, it was inevitable that the binary code used inside the machine would be reduced to a simple logic. That this simple logic has now been transformed into hardware is very largely due to the emergence of integrated circuits. The binary code consists of counting in powers of 2. The decimal number 21 is $2^4 + 2^2 + 2^0$, or in binary notation 10101. Similarly, 1023 is $2^9 + 2^8 + 2^7 + 2^6 + 2^5 + 2^4 + 2^3 + 2^2 + 2^1 + 2^0$ or 1111111111 in binary notation. All digits are either 1 or 0, so any two-state device which can be triggered from one state to the other is the fundamental building brick for binary counting.

A simple example of this is the RTL flip-flop, where RTL stands for ' resistor-transistor-logic '. The circuit is shown in fig. 2.3.

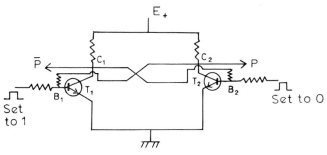

Figure 2.3.

The action of this circuit is as follows: If a positive pulse is applied to the base of the transistor $T_1$, it will cause current to flow (if it is not already flowing) in this transistor. This drops the potential at $C_1$ and the potential on the base of $T_2$ which remains unconducting or ceases to conduct. When the pulse is removed from $B_1$ the situation remains the same—$T_1$ conducting and $T_2$ cut off.

If a similar positive pulse is applied to $T_2$ the opposite effect occurs and $T_2$ conducts while $T_1$ is cut off. The first pulse may be designated a ' set to 1 ' pulse and the second a ' set to 0 ' pulse and after either the flip-flop is said to *store* either a 1 or a 0. Signals can be taken off at $C_1$ and $C_2$. These signals are the ' inverse ' of each other, i.e. if $P$ is 1, $\bar{P}$ is 0 and if $P$ is 0, $\bar{P}$ is 1.

It may be seen from fig. 2.3 that the flip-flop is a combination of two exactly similar circuits or ' modules ' connected by two feedback loops. The combination is called an O–I module, or alternatively a NOR module, meaning that it functions as an OR gate followed by an ' inversion ' or ' negation '. An OR gate is one in which if any input is 1, the output is 1. For example, if three inputs of positive potentials

21

1, 1, 0 were applied the output would be 1. Similarly, if inputs 1, 0, 1 were applied the output would be 1. Only in the case where the inputs were 0, 0, 0 would the output be 0.

The O–I module inverts the output signal for, as can be seen by referring to fig. 2.3, a positive pulse at the base of $T_1$ results in a drop in potential at $C_1$. So finally we have 0 output for all the cases mentioned in the last paragraph except where the inputs are 0, 0, 0. Then the output is 1.

Bearing these results in mind, it is possible to examine how a binary full adder operates, using O–I modules only. The input pulses will be labelled A, B, C, but it must be remembered that these represent either a 1 or a 0. The 1 is a positive pulse and the 0 is an absence of pulse. Adding 1 to 1 to 1 in the binary code gives the binary number 11, i.e. $2^1 + 2^0$, whereas adding 1 to 1 gives the binary number 10. The first circuit (fig. 2.4) gives the carry in the addition, for example,

$$1 + 1 + 1 = 1(\text{carry}) \ 1(\text{sum}),$$

and this may be readily checked by taking various values of A, B and C. The case of A = 1, B = 1, C = 1 has been indicated by the digits in

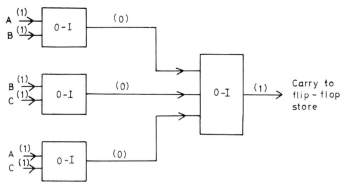

Figure 2.4.

brackets. The first three modules have two inputs, or, as it is expressed in computer language, there is a ' fan-in ' of 2. At the output O–I there is a fan-in of 3. All input signals are in parallel. The final digit in the addition must now be found and the full circuit for the adder is as shown in fig. 2.5 where the part inside the broken line is fig. 2.4.

Again it may be readily checked that the circuit adds correctly. If four or more additions are to be performed in parallel many more modules are required and it is obviously a great advantage from a manufacturing point of view if all these modules are as nearly as possible the same.

This is why integrated circuits have become so important. Using modern techniques the whole of the adder shown diagrammatically in fig. 2.5 can be accommodated on a single crystal—a so-called monolithic circuit-and the power drain is diminished accordingly. Consequently, modern computers can accommodate a very large arithmetical function.

The O–I module is an example of RTL or resistor transistor logic and

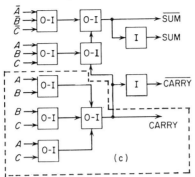

*Figures 2.4 and 2.5 are reproduced by permission of D. Van Nostrand Co., Inc., Princeton, N.J., from the book entitled "Electronic Digital Components & Circuits" (1967).*

Figure 2.5.

in all there are three combinatory procedures used in any binary transistor logic. The first is the operation OR which, as has already been mentioned, means the combination of two or more signals so that either or any one of them gives an output. I.e. $1 \oplus 1 = 1$, $1 \oplus 0 = 1$, $0 \oplus 0 = 0$. The second is the operation AND which means the combination of two or more signals so that both or all yield an output. The symbol for this operation varies considerably. If the symbol $*$ is used, AND for two inputs gives $1 * 1 = 1$ and $1 * 0 = 0$. The third is the operation NOT which simply means that a 1 becomes a 0 and a 0 becomes a 1. NOT is symbolized by $\ominus$. These procedures are based on an algebra developed by George Boole in 1847 and called Boolean algebra or ' symbolic logic '. If we replace our arithmetical values 1 and 0 by A and B, Boolean algebra gives the following laws.

An OR operation is symbolized by $C = A \oplus B$.

An AND operation is symbolized by $C = A * B$.

A NOT operation is symbolized by $C = \bar{A} = \text{not } A$.

It can be seen that all three operations have much greater force where A and B are 1 or 0, for then hardware can be constructed to do these operations electronically.

5. *The applications of transistor logic*

A brief explanation of the functioning of a particular logic module, the O–I element, has led to the description of a full adder, i.e. an adder

23

which will give the sum and the carry for any two binary digits. This is a first step towards producing an adding machine, but, as will be seen immediately, once this step is understood the organization of similar adding units to form a machine is a comparatively simple matter. Figure 2.6 shows diagrammatically how it is done.

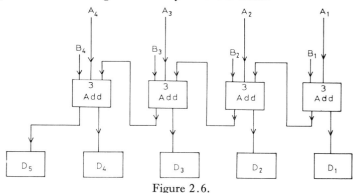

Figure 2.6.

The figure shows an adding machine capable of adding two 4-digit binary numbers, $A_4A_3A_2A_1$ and $B_4B_3B_2B_1$. Signals representing these numbers are applied to the full adders, the normal symbol for these being as shown. $A_1$ and $B_1$ give the sum $D_1$ and a carry to the second adder. This accepts $A_2$ and $B_2$, giving $D_2$ and a carry to the third and so on, until the two outputs from the 4th adder are $D_4$ and $D_5$, $D_5$ representing the carry. This is the fastest type of adding machine because all the operations are carried out in parallel.

Similar transistor logic can be applied to subtraction, multiplication and division, and by means of a carefully programmed procedure to the addition or multiplication of the most complicated mathematical functions.

Enough has been said, however, to convince the reader of the importance of transistor logic; we have now to show how frequency conversion comes into this field at all.

Modern computers do their arithmetic at fantastic speeds, but they are actuated by pulses. To achieve these high speeds it is essential to produce pulses which have the minimum ' rise-time '. The latter term is used to describe the time taken by the pulse voltage to rise from zero to its designed height. Figure 2.7 may help to make the concept clear.

The figure is a graph of the voltage of a 2 microsecond pulse (2 $\mu$s) against time. It will be noted that the rise-time, $T$, is about one quarter of a microsecond, so, if this pulse were used to actuate a computer, the latter would be dead for $\frac{1}{4}$ $\mu$s after every pulse was applied. Moreover, in proceeding through the various logic devices it would tend to become more rounded, if nothing were done about it, finishing up as shown by the dotted graph.

24

Consequently pulses are very carefully shaped before being injected into the machine and pulse-shaping circuits are also introduced wherever necessary along the paths of pulses. These rejuvenate the pulses, not only reducing the rise-time but increasing the amplitude simultaneously. The pulse shown in fig. 2.7 may, so far as the computer is concerned,

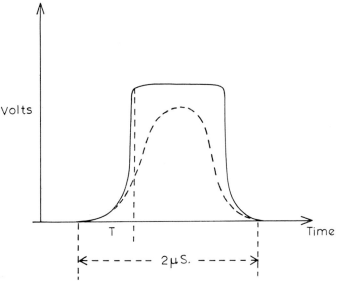

Figure 2.7.

represent two steady state conditions, one of which persists for 2 $\mu$s. Fully pulsed computers are much less common than they were but the need for rapid rise-time to a new semi-permanent value of the voltage makes shaping circuits a necessity still. These rapid rise-times can be regarded as frequency conversions, particularly when they are obtained from sinusoidal oscillations for the following reason:

Consider the sine-wave shown in fig. 2.8, and imagine that from this sine-wave pulses A, B, ... are formed. Let the wave be represented by $\sin 2\pi f t$, then the pulse repetition frequency is $f$. Let the width of the pulse be $w$, i.e. $w =$ the time in seconds during which the pulse exists. Also let it be assumed that each pulse is a perfect rectangle.

Then analysing the pulses, using the Fourier series, and calling the pulse disturbance $p(t)$:

$$p(t) = wf + \frac{2}{\pi} \sum_{n=1}^{\infty} \frac{\sin (n\pi wf)}{n} \cos (2\pi n f t),$$

where $n$ is an integer. The cosine waves making up the rectangular pulse extend from frequency $f$ to frequency $nf$, where $n$ is any integer. But the amplitude of the wave frequency $nf$ is $2 \sin n\pi wf / \pi n$, and so as $n$

25

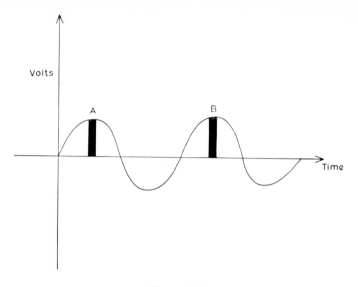

Figure 2.8.

increases the amplitude decreases. Clearly the effect of the wave having $n = 100$ is negligible under all circumstances, and so the circuits carrying the pulses should have a band-width of about 10–50$f$ if a good rectangular pulse is to be obtained. So, once more we find that frequency changing is an essential element of computer technology.

Indeed in any case where the wave-form of the original voltage variation is changed to a different wave-form, the essential element of frequency changing occurs, although in the majority of practical cases it results in harmonics of the original periodicity being produced, accentuated or diminished.

# CHAPTER 1.3

### transducers

THE concept of a transducer in its modern form is not very old. A transducer is any device which transforms energy in one form into energy of another form and this concept is so wide that it covers such diverse devices as a coal fire, a nuclear pile, a piezo-electric crystal, a thermocouple or a microphone.

The concept can be broken down into two distinct classes of transducer. The first and most important comprises devices which we use as energy convertors on a large scale. The fire and the nuclear pile would fall into this category. The second class comprises transducers such as the thermocouple or (in some cases) the microphone. Here the energy transformation is used to measure some effect. Such transducers are now known as ' instrument transducers ' and in this second class the most important species is that which transforms some other form of energy into electrical energy. Table 1.3.1 is taken from a paper by one of the authors to the *Journal of Scientific Instruments*. It lists all possible energy transformations and gives a number of examples of transducers which produce these. The Institute of Physics and The Physical Society are thanked for permission to use it.

If the concept is extended slightly, all *active* electronic devices such as electronic tubes, crystal diodes and transistors can be termed electronic transducers where the transformation is not from another form of energy to electrical energy but from one form of electrical energy to another. For example, an electronic amplifier is actually a black box which transforms d.c. energy into energy at some required frequency or frequencies. In strict scientific terms the electronic tube is not an amplifier. It is a device which takes some of the d.c. energy from its power supply and turns it into energy at the signal frequency. The transistor can be described in the same way.

All the frequency transformations which have been discussed in earlier chapters have required the use of transducers to bring them about and, as will be seen later, the same concept can be applied to many other transformations.

It is perhaps worth while at this stage to move away from electronic devices to the older technologies of mechanical and electrical engineering with a view to examining how nearly frequency transformations by mechanical or electrical means correspond to the electronic processes dealt with in the previous chapters.

The simplest mechanical method of frequency division is a train of

Conversion from

| Types of energy | Nuclear | Atomic | Elastic | Heat | Static electromagnetic | Radiant electromagnetic | Microkinetic | Macroscopic kinetic |
|---|---|---|---|---|---|---|---|---|
| Nuclear | | Direct action is impossible | Direct action is impossible | Above $10^5$ degrees K | Not entirely impossible | Transformations by means of gamma rays | Transformations by means of particles | Impossible in laboratory |
| Atomic | Radioactive transformations (a by-product) | | Direct action is impossible | Thermal combination, dissociation and excitation Endothermic reaction | Stark and Zeeman effects Electric discharge | Photo-chemical reactions | Chemical change by particle bombardment | Detonation by impact(?) |
| Elastic | No apparent direct action | Crystal formation | | Crystal formation | Piezo-electricity Ferro-electricity Magneto-striction Loud-speaker | Modifications to molecular structure | Modifications to crystal structure | Generation of elastic oscillations |
| Heat | No direct action, but final form | Exothermic reaction (excess) | Attenuation and supersonic heating | | Resistive losses | Absorption and degradation | Bombardment of matter in bulk | Friction and viscosity |
| Static electro-magnetic | May occur but unknown | Volta effect | Piezo-electricity Ferro-electricity Magneto-striction Microphone | Peltier effect | | Radio-frequency reception | Special high tension supplies | Dynamo, etc. |
| Radiant electro-magnetic | Gamma radiation | Chemical excitation of radiation | Feasible by means of the three effects above | Radiation from hot body | Radio-frequency emission | | X-ray tube Alpha- or beta-ray transformations with gamma emission | Possible perhaps in future |
| Micro-kinetic | Alpha and beta particles Neutrons Mesons | Chemical ionization | An effect should be possible | Thermionic emission Molecular beam | Field emission Hall effect Particle accelerators | Photo emission Gamma-ray transformation with alpha and beta emission | | Only via chemical action |
| Macro-scopic kinetic | Final product in nuclear engine | Final product in i.c. engine | Spring motor or elastic motor | Steam engine | Electric motor | Radiation pressure effects | Practical applications feasible | |

Conversion to

gears. Suppose the train begins with a ten-tooth pinion meshing into a hundred-tooth pinion. On the same spindle as the hundred-tooth pinion let there be a ten-tooth pinion which again meshes into a hundred-tooth pinion and let the procedure be repeated as many times as are necessary. If now the angular velocity of the first spindle is 100 rev./min, the angular velocity of the second is 10 rev./min, of the third 1 rev./min, and so on, and since angular velocity has the same dimensions as frequency (time$^{-1}$), this chain represents a frequency division of 10 per pair of spindles or of $10^{n-1}$, where $n$ is the number of spindles.

This machine, made up of spindles and pinions, consists of purely passive elements. They are made active by being rotated by a prime mover which may be an electric motor or a piston engine or even a gas turbine. The combination of the prime mover and the machine *may* be termed a transducer although the transformation is only in angular velocity and the entire device is probably only a small part of a piece of test equipment for measuring the input and output energy of the prime mover.

The analogous electronic device has been described in an earlier chapter. It is the ' dekatron ' or a series of flip-flops connected as a ring counter which performs exactly the same operation. In a large number of cases there is a mechanical analogue to an electronic device and the example which is about to be cited is of this nature.

If a mechanical shock is administered to a mechanical resonator, such as a bell or any crystal, the resonator will ring. Due to air friction, friction at supports and in the metal, the ringing tone gradually diminishes in amplitude. The mechanical damping may, however, be made small, and so a bell may be maintained as a continuous oscillator by striking it with the clapper once every few seconds. The shock given to the bell may be represented by the pulse OAB in fig. 3.1 *a*, while the

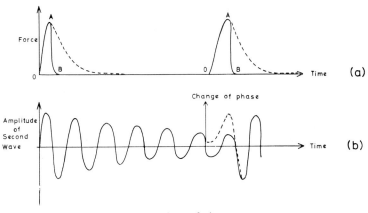

Figure 3.1

29

sound wave generated may be represented by the damped wave shown in fig. 3.1 *b*, giving a tone at a frequency *f*, associated with harmonics and in many cases with other related tones. This is the analogue to the frequency multiplier already described (p. 6) where the signal frequency was $f_0$ and the transistor or electronic tube was biased to provide a small spike once per period. This voltage spike affected a parallel resonant circuit at frequency *nf*, the resonant frequency of the circuit where *n* is an integer, and gave a damped sinusoidal oscillation at that frequency. The only important difference is that there is no abrupt change of phase in the electronic circuit, because of the relation between the two frequencies, whereas in the case of the bell this change of phase occurs and gives the extraordinary pattern of sound we obtain in campanology. Moreover, in the multiplier the attempt is made to construct a circuit having only one degree of freedom, whereas in the bell the resonator has many degrees of freedom and therefore many resonant frequencies. This is another example of mechanical frequency conversion.

## 1. *Unwanted vibrations*

The example just given leads naturally to a subject which is becoming more important every day. The nuisance value of this type of conversion is very high and in many cases it contributes to the unreliability of ill-designed components. The nuisance value will be considered first.

Anyone who uses a motor-car is aware of the trouble which can be caused by vibration. This vibration is often at the mechanical resonant frequency of some component. A good example arises from the lack of balance in the wheels of the motor-car. Where this exists, no matter how small it is, when the car is travelling at high speed the steering column starts to vibrate quite appreciably. This vibration is transmitted to the hands and arms of the driver and, apart from the possibility of danger, can be very tiring. The reason for this vibration is that the unbalance in the wheels, imperceptible at low speeds, reaches a point where the frequency of the impulses is sufficient to set up sympathetic vibrations in many of the structures making up the motor-car. If the steering column is assumed to be effectively 5 ft long its fundamental mode of vibration is at a wave-length of 10 ft, corresponding roughly to a frequency of 110 Hz in air at N.T.P.

If now the wheel diameter is 2 ft, its circumference is 44/7 ft and when the car is travelling at 60 m.p.h., or 88 ft/sec, each wheel makes a complete revolution in 1/14 sec. Assuming that all four wheels are unbalanced, this means 56 impulses/sec, but these would not normally be equally spaced and so the wheel impulses—at the rate of 50 to 60/sec. would set up a sympathetic vibration in the steering column.

Another example of the same type arises from engine revolutions. When these reach about 100 c/s or 6000 r.p.m. they may set up the rumbling roar which is the characteristic of a badly designed chassis—in this case the air in the body cavity of the car resonates to the engine vibrations.

Aircraft exhibit similar phenomena. In the days of piston engines, vibration of the wings was a matter for concern. Today the high-pitched whine of the jet-engine is only heard on the ground or when landing. The greater part of the noise heard by passengers is due to small variations in the pressure and temperature of the air through which they pass and even this noise virtually disappears at 20 000 ft. The vibrations, however, are still present, although they are not heard and the ageing of steel and light alloy structures in aircraft is largely due to the continual vibration sympathetically produced by the engines.

Turning now to reliability, resonances in components which have to operate in the vicinity of any vibrator can be very serious. Only one example will be quoted—the common electronic tube used in a radio receiver.

Needless to say, electronic tubes can fail for a number of reasons, many of which are the direct result of operating the tube outside the maker's specification. Vibration is, however, the only cause to be considered here. A typical Service Specification calls for the tube to stand up to 10–500 Hz at 5 g and 50–2000 Hz at 10 g. These figures mean that the maximum accelerations which the tubes must stand are 5 g up to 500 Hz and 10 g between 50 and 2000 Hz. A quick calculation shows that the amplitude of oscillation at 2000 Hz to give 10 g is only about $6 \times 10^{-7}$ m.

Nevertheless, this is quite sufficient to cause measurable resonances if any of the members making up the tube's structure resonate at one of the frequencies. The examination can be done stroboscopically and the chief resonances identified by the response at several frequencies.

Aircraft are still one of the major users of electronic tubes and aircraft are notoriously subject to vibration. Nevertheless, with good design and careful manufacture such tubes can be made to give lives of 10 000 hours on the average. All components on an aircraft have the same stringent conditions to pass. Indeed in the case of components mounted on the engines themselves the conditions are much more severe. One of the important advantages of using transistors instead of tubes is that the transistors have fewer resonant frequencies at higher values.

## 2. *Optical frequency conversion*

Any interferometric arrangement produces a frequency conversion which becomes easily apparent if the fringes produced by interference between the two beams of monochromatic light are counted along a spatial axis at right angles to the fringes.

For example, suppose it is necessary to measure the refractive index of a gas at atmospheric pressure. This can be done by splitting a light beam by the method of Jamin or Raleigh or Michelson. Since the last-named interferometer is best known, the measurement will be described in terms of its use. A diagrammatic representation of one form of interferometer is shown in fig. 3.2.

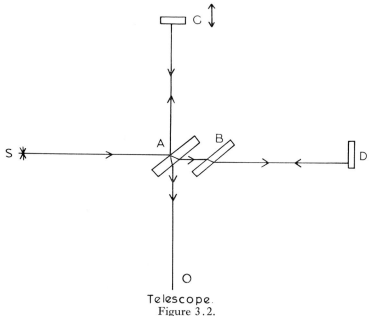

Telescope.
Figure 3.2.

The light from a monochromatic source falls on the parallel-sided plate A where one beam is reflected to the silvered plane mirror C and the other proceeds through a similar parallel-sided plate B to the plane silvered mirror D. The back surface of A is semi-transparent so that the two beams, that towards C and that towards D are equal in intensity. The beam reflected by C returns through A to the telescope at O, while the beam reflected by D returns through B to the telescope. It is possible, therefore, by adjusting the position of C as shown by the arrows at the top of the figure to make the optical paths of the two beams almost exactly equal.

Without going into too much detail, it should be mentioned that all precautions are taken to ensure that the paths can be made identical and that any movement of C is at right angles to the beam.

Then if a tube connected to a high vacuum pump is placed in one path, say between B and D, the instrument can be adjusted so that the effect of the glass ends to the tube is nullified by moving C away from A. The fringes appear when the path differences are nearly identical. The tube is completely evacuated and the necessary adjustment is made to C. The tube is then slowly filled with the gas whose refractive index is required. Fringes pass across the telescope and are counted. The number of fringes which pass across the eyepiece of the telescope multiplied by the wave-length of the monochromatic light gives the path difference due to the gas, the refractive index of which can now be calculated. The counting can be done using a photocell as a detector and connecting it to a counter. In this case the frequency of the optical wave (about $5 \times 10^{14}$ Hz) has been brought down by the interferometer to a frequency of a few Hz, assuming that the counting of the fringes takes a few seconds.

It should be noted that interference between two systems of waves, whether they be mechanical waves, hydromechanical waves, sound waves or light waves is of an entirely different nature from the modulation or heterodyning of electric currents in circuits containing the active elements discussed in Chapter 1.

It will be remembered that to produce frequency conversion by means of an electronic tube or a transistor, a curved V—I characteristic was required. But an interference pattern is produced by two moving waves wherever they superimpose. The simple theory of this kind of frequency conversion assumes that both waves have the same frequency and amplitude, but have different phases. A sinusoidal plane travelling wave can be represented by the expression:

$$A \sin 2\pi \left(\frac{t}{\tau} - \frac{x}{\lambda}\right)$$

or

$$A \sin 2\pi \left(ft - \frac{x}{\lambda}\right),$$

where $\tau$ is the period of the wave, $f$ is the frequency, $A$ is the amplitude, $\lambda$ is the wavelength and $x$ is the distance measured in the direction in which the wave is travelling.

This representation holds for any sinusoidal travelling wave no matter what its nature may be.

To obtain interference let this wave be superposed on another, differing only from the first in phase. Let the second wave be represented by $A \sin 2\pi (ft - (x+a)/\lambda)$. Then the two waves together will give a disturbance:

$$D = A \left[\sin 2\pi \left(ft - \frac{x}{\lambda}\right) + \sin 2\pi \left(ft - \frac{x+a}{\lambda}\right)\right]$$

and, since

$$\sin x + \sin y = 2 \sin \left(\frac{x+y}{2}\right) \cos \left(\frac{x-y}{2}\right),$$

$$D = 2A \sin 2\pi \left(ft - \frac{x+a/2}{\lambda}\right) \cos \frac{\pi a}{\lambda}. \qquad (3.1)$$

$D$ is a wave having the same frequency as its two components but an amplitude of $2A \cos (\pi a/\lambda)$. But $\cos (\pi a/\lambda)$ can vary from $-1$ to $+1$ and therefore the amplitude can vary from $-2A$ to $+2A$. Where cos $(\pi a/\lambda)=0$, there is no disturbance and where cos $(\pi a/\lambda)=-1$ or $+1$ there is a maximum disturbance.

Setting up the interferometer gives a set of bright lines in the telescope. Any change in the length of one of the optical paths will cause a change in the pattern of lines, since this alters $a$. But the phenomenon of interference owes nothing to non-linearity. In fact it is a purely linear action. Only one frequency (the original) is produced with a variable amplitude, but the variable amplitude gives rise to a spatial distribution which can be transformed into a low frequency variation by changing one of the optical paths.

The only other example of frequency conversion of practical importance is the Doppler effect, more correctly designated the Doppler–Fizeau effect. It is well known. If a source of any type of wave-motion is travelling towards a stationary observer, the frequency is higher than it would be if the source were stationary. Similarly, if the source is travelling away from the observer the frequency is lower. If, on the other hand, the source is stationary and the observer is travelling towards it or away from it, the frequency is increased or decreased in much the same way. The magnitude of the frequency shift can be calculated as follows:

(a) Source moving towards the observer—velocity of wave, $V$, velocity of source, $v$, wavelength and frequency when both source and observer are at rest, $\lambda_0$ and $f_0$ respectively.

When the source is moving towards the observer the wavelength is shortened. In one period the source approaches the observer by an amount $v/f_0$.

Therefore the wavelength to the observer is $\lambda_T$, where

$$\lambda_{T_1}=\frac{V-v}{f_0},$$

or, since $\lambda_0=V/f_0$,

$$\lambda_{T_1}=\left(\frac{V-v}{V}\right)\lambda_0. \tag{3.2}$$

(b) Source moving away from the observer—in this case $v$ in the above equation is negative and if the speed of recession of the source is $v$:

$$\lambda_{T_1}=\left(\frac{V+v}{V}\right)\lambda_0. \tag{3.3}$$

(c) Observer moving towards a stationary source—using the same symbols, the observer will meet $v/\lambda_0$ more waves per second. He therefore receives $(V+v)/\lambda_0$ wavelengths per second and the observed wavelength $\lambda_{T_2}$ is given by:

$$\lambda_{T_2}=\frac{V}{f_{T_2}}=\left(\frac{V}{V+v}\right)\lambda_0. \tag{3.4}$$

34

(*d*) If the observer is moving away from the source, the sign of $v$ changes and

$$\lambda_{T_2} = \left(\frac{V}{V-v}\right) \lambda_0. \qquad (3.5)$$

It should be noted that if $v$ is very small compared with $V$, then the formulae for the source moving correspond closely to the formulae for the observer moving, since

$$\left(\frac{V}{V+v}\right) \lambda_0 = \frac{\lambda_0}{(1+v/V)} \doteqdot \lambda_0(1-v/V)$$

when $v$ is small. But

$$\lambda_0(1-v/V) = \left(\frac{V-v}{V}\right) \lambda_0,$$

making the formulae identical so far as first-order terms are concerned.

A slight warning should be given here. The calculations just concluded are based on Newtonian mechanics. If relativistic mechanics are employed, the formulae are slightly different. For example, equation (3.4) becomes:

$$\lambda_{T_2} = \left(\frac{V}{V+v}\right) \lambda_0 \sqrt{(1-v^2/V^2)}, \qquad (3.6)$$

which is the exact expression for an optimal Doppler shift. But this only holds for $V = c =$ velocity of light and in any case equations (3.4) and (3.6) agree to the second order of $(v/V)$.

The Doppler–Fizeau effect was first used in astronomy, but is most easily observed in acoustics. A locomotive emitting a whistle of fixed frequency and passing under a bridge on which the observer is standing appears to give rise to a higher note as it approaches and to a lower note as it recedes. This effect may have been put to good use since the end of World War II in what is known as Doppler radar, the principle of which will now be explained.

The unsophisticated radar used in World War II merely gave the range and position of an attacking vehicle, usually an aeroplane. With the coming of I.B.M.'s and other rockets, it became necessary to identify not only the range and position of the object, but also to obtain as quickly as possible a good plot of its direction of motion. If time were not important, this could be done using two or more radar stations and computing the course from a series of observations. But time is of vital importance and so the alternative method—Doppler radar—can be used. Each transmitter frequency is stable over the time required for a pulse of energy to be transmitted, reflected and received. The received signal will differ in frequency from the transmitted signal by an amount depending upon the component of the target's speed towards or away from the tracking station. Theoretically, therefore, two such stations, the requisite distance apart, could compute from a single observation the course of the target. In practice more than two stations could be used and very often more than one observation would be taken. This could be the most important application of the Doppler–Fizeau principle to be used to date.

Clearly there are many modifications of such a system for use with tactical weapons, but enough has been said to indicate the field of endeavour. It is interesting to calculate the magnitude of the Doppler change of frequency in the case just discussed. Suppose the target is moving towards the radar station with a component of speed $v$. Then the received wavelength $\lambda_R$ in terms of the transmitted wavelength $\lambda_T$ is given by:

$$\lambda_R = \frac{\lambda_T(1-v/c)}{(1+v/c)} \fallingdotseq \lambda_T\left(1-\frac{2v}{c}\right), \tag{3.7}$$

$$f_R \fallingdotseq f_T\left(1+\frac{2v}{c}\right). \tag{3.8}$$

Let $v = 3000$ km/s and $\lambda_T = 3000$ MHz, then the change in frequency is 2 parts in $10^4$ or 600 kHz. The components of the velocity of the target towards two (or more) stations can then be deduced and by vector addition the course of the target can be obtained.

Another possible application of the Doppler–Fizeau principle is to road safety. Suppose that all vehicles travelling on motor-ways were fitted with a crystal-controlled ultrasonic transmitter which beamed energy backwards in the direction opposite to the direction of travel. Suppose also that all such vehicles were fitted with a simple receiver to pick up the radiations from vehicles travelling in front of it. Then, if the transmitted frequency were in the neighbourhood of 30 kHz and the receiver used some of the energy of the transmitter frequency as its local oscillator, two cars which were closing at the rate of 60 m.p.h. would give rise to a high pitched whistle in the car at the rear. The pitch of the whistle would gradually decrease as the two speeds became equal, so that, even in fog, no vehicle could overtake any other vehicle. The discomfort caused by a very intense low-frequency note would warn the on-coming driver that he was approaching too closely the traffic ahead of him.

If the reader cares to work out this problem in detail, he or she will find that the proposal, if adopted, would not affect other road users and could substantially reduce the accident rate on motorways, particularly during fog.

Frequency transformation is indeed a useful tool in this age of technological change.

# FREQUENCY CONVERSION

## Light Conversion Devices    Part II

**W. E. Turk**
**English Electric Valve Co. Ltd.**

*Acknowledgments*

The author gratefully acknowledges the assistance derived from many published and unpublished sources in writing this part of the book. It is regretfully impossible to mention all of these but an attempt has been made to include those which could lead to further reading.

For diagrams, resort has been made to manufacturers' data sheets and these are acknowledged individually.

Particular thanks are due to Mr. A. J. Young, Managing Director of English Electric Valve Company, who sponsored the work and permitted much of the Company's vacuum tube processing to be included.

The author also wishes to thank his several friends for valuable help in the preparation of the manuscript. He emphasises, however, that all views and opinions expressed are his own and not necessarily those of English Electric Valve Company.

# FOREWORD

BECOMING of greater importance in everyday life is that science or art concerned with the conversion of light into electrical currents and vice versa. This section of the book deals with devices involving a conversion process in which light is either absorbed or emitted. The wavelengths involved are those normally defined as the optical region. Within this range it is convenient, and unavoidable, to consider sub-divisions of fairly narrow band-width. The reason for this is that no material is photosensitive over the whole region and, likewise, no light emitter does so with a spectrum over the entire optical band.

The broad compass of the main title, allows other conversion processes such as photo-synthesis, electro-thermal radiation, incandescence, X-ray detection, lasers, etc., to be legitimately included. However, space considerations prevent this and only electro-optic interactions will be discussed. Furthermore, those processes which involve a visual display of thermal and acoustical images will be mentioned only briefly. Their importance to medicine necessitates this mention.

Various terms, such as ' opto-electronics ', are in vogue to define the subject, but none adequately condenses the phrase ' light conversion devices '. Confusion with the well established ' electron-optics ' must be avoided. This latter term refers only to the design of electron deflecting fields, either electrostatic, magnetic or combinations of the two which occur in vacuum tubes where electron beams are controlled in the same manner as are light rays in physical or geometrical optics.

There are few means of converting light into electrons. Each is basically a photo-cell. The reverse process of producing light by means of electrons begins with the simple incandescent lamp bulb, includes the neon tube and electron excited phosphor and culminates in the light emitting crystal of which a typical example is gallium arsenide.

# INTRODUCTION

THE basic process of converting light into electrons requires a photo-electric transducer. The conversion process is made manifest in one of three ways, the liberation of free electrons into either an electrolyte, a crystal lattice or a vacuum. These effects, given in order of their historical discovery, are respectively the photo-voltaic effect[1], the photo-conductive effect[2] and the photo-emissive effect[3].

Strictly speaking, only devices depending wholly on these simple effects come within the scope of the present work and they will be adequately described. However, their application to, and inclusion in, more complicated structures is felt to justify an extended treatment of the latter. The student entering the realm of industrial photo-electronics will be less concerned with the basic photo-electric element than with its role in a more sophisticated and demanding unit—itself, in turn, merely a building block in a complicated piece of equipment.

In discussing the production of light by electrons, the heating effect of an electric current, incandescence, will not be considered. Only the emission of light caused by electron movements between atomic energy levels is within the scope of this present work.

# CHAPTER 2.1

## photo-electricity

*Theoretical considerations*

For the detailed theory, the student is referred to the recognized works where a comprehensive analysis of the subject may be found[4]. A complete treatment involves mathematics of a level above that of the students for whom this work is intended, but familiarity with ONC and Intermediate Degree Standard is assumed. This chapter is aimed at refreshing the memory and will introduce the several practical concepts which are involved.

To explain the interaction of light and matter resort is conveniently made to the Bohr atom[5] in order to construct a working model with which the creation of free electrons can be explained.

Basically, the theory attributes to the atom a structure consisting of a central nucleus around which move orbital electrons in discrete paths and in definite numbers. The nucleus contains protons and neutrons which determine the atomic weight of the element and also the total number of electrons in the outer orbits or shells.

Each shell can contain only a certain number of electrons and as the atomic weight increases so inner shells become filled and more electrons are found in the larger orbits more remote from the nucleus. The binding forces between electron and nucleus decrease as the orbital radius increases.

All materials of course, consist of a conglomeration of atoms, and do not behave as single atoms. Each atom is influenced by its neighbours and the behaviour of the atomic lattice as a whole must be considered. The lattice may be orderly as in a crystal, or random as for amorphous materials. In any system, due to the interaction between atoms, the individual electron orbits are displaced and give rise to bands of quantum states or energy levels. For the innermost electrons the atomic interaction is small and the energy bands are narrow. On the other hand, for the outermost electrons—those determining the atomic valency—the grouped energy bands cover a range comparable with that of the energies of the electrons and some overlapping of different shells tends to occur. The distribution of electrons in these bands conforms to a strict statistical pattern and one can evolve a definite probability function to indicate the existence of an electron at a particular level. The energy level at which this probability is 50% is termed the Fermi level. Above the valency band, which is completely filled, exists an energy level which is rarely filled and from which electrons can easily be removed to allow conduction. This level is termed the 'conduction band' and the gap between it and the valency band is often referred to as the 'forbidden gap'.

For a material to be a conductor, i.e. to give rise to a flow of electrons, there must be electrons present in the conduction band. For metals this is a natural state the electrons having moved from the valency band to the conduction band by reason of their thermal energy. For other materials the magnitude of the band gap determines the ease with which valency electrons can be raised to the conduction band. If the gap is greater than 2 eV the material is classified as an insulator† while those possessing smaller band gaps are termed semiconductors. In these latter, the conducting state may be created by thermal excitation of the electrons to cross the forbidden gap. The 'holes' created in the valency band by the lost electrons allow an electron movement within it if an external field be applied. This gives rise to further conduction— 'hole' conduction.

A further variation of the conditions occurs in materials containing minute amounts of impurity—just enough, in fact, to upset the otherwise regular atomic arrangement. Normally, in insulators with band gaps of greater than 2 eV, the presence of the impurity produces a new quantum level within the forbidden gap, so reducing it and facilitating electron transfer. Two types of impurity effect can occur. One, in which the new quantum level occurs near the top of the forbidden gap, i.e. near to the conduction band. In this case electrons can be raised from it to the conduction band to allow conduction. In the second case the new level occurs near the bottom of the forbidden gap when electrons can be excited to it from the filled valency band and conduction occurs by means of the 'holes' created therein. These two types of conduction are termed respectively n type and p type.

In summary then, conduction in a material—i.e. the passage of an electron current following the application of a stimulating electric field— can occur either when electrons exist in the atomic conduction band or 'holes' exist in the valence band. For either of these conditions to be created in a material, normally an insulator, energy has to be supplied. This excitation may be thermal or, as is the interest of the present treatment, by photons. The photon being the unit or 'atom' of light.

In the case of the photovoltaic effect, free electrons are not created to form a current under the influence of an applied field. Instead, a circuit, external to the transducer is energized by the transducer itself as a result of an internally generated e.m.f. The effect can be explained by similar reasoning to that used earlier.

A photovoltaic cell, either of solid or of solid and liquid materials, consists essentially of a junction between two dissimilar substances. If it is supposed that each of the materials is at earth potential before they make contact then, since the Fermi levels in the two are not identical,

† See Part I.

more electrons move in one direction than in the other. This creates a potential difference at the junction which is just sufficient to prevent the flow of electrons. Indeed, if the metals are now insulated from earth, the so-called contact difference of potential (difference in Fermi level potential) can be measured. When illuminated, electron/hole pairs are created in the neighbourhood of the boundary and a flow of current will occur across it and can be measured in an external circuit—the device behaving as a battery, since the balance between the Fermi levels has now been disturbed.

Unlike the physical effects it describes, the foregoing theory is all comparatively recent in formulation. For example, the photovoltaic effect was discovered by Becquerel in 1839 but no credible explanation arrived for nearly a hundred years.

This is not to say that photoemission has not been studied during the intervening period. The experimental results obtained by many workers have all contributed towards establishing facts for which explanations have been required.

Investigational work is hampered by the inherent difficulty of obtaining measurable photo-effects from simple and fully understood materials. In the basic case, except in the extreme ultra-violet, no photon/atomic reaction can be detected with hydrogen—the simplest atom. In contrast, the emission spectra of atomic hydrogen have been studied, calculated and demonstrated in magnificent detail[6].

Since the energy of a photon is low, only those materials with easily accessible electrons are readily investigated. This would indicate that molecules of high atomic number, i.e. with an outer shell of lowest energy and containing the minimum number of electrons, would be most easily activated. By reference to the periodic table of the elements it can easily be understood why caesium is a likely contender as an efficient photo-active material. It does, in fact, form the basis of all highly sensitive photo-emitters.

For the photo-conductive and photo-voltaic effects no simple mechanical atomic model can be postulated. Research into these two effects illustrates how painstaking ingenuity and intuition yields positive results from a purely empirical approach.

For photo-emitters the following experimental facts are given by Sommer[7]:

(1) Photo-emission can only occur when the wavelength of the incident light is below a certain value.

(2) The energy of the photo-electrons differs according to the material.

(3) The energy of the photo-electrons is independent of the intensity of the incident light.

(4) The number of the photo-electrons is linearly proportional to the intensity of the incident light.

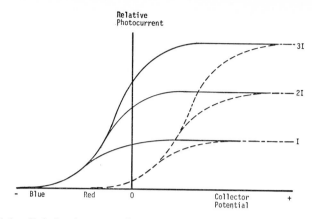

Fig. 1.1. Relation between photo-current and collector potential for blue and red light in a simple photo-cell.

We may present these results graphically (see fig. 1.1).

Postulating a certain monochromatic red, then the three curves shown would be obtained with intensities of *I*, 2*I* and 3*I* of red and blue light incident upon a given photo-cathode. The intercept on the potential axis will differ according to the photo-cathode material and will also become more negative as the wavelength of the illumination decreases— i.e. as the photon energy increases. An experiment to illustrate this latter fact will also show that the number of photo-electrons emitted varies according to the photon energy. The relationship is not linear as one might expect it to be and a completely acceptable explanation of the phenomena has yet to be formulated. The optical filtering effect of the photo-cathode and the precise position of the emitting layer within the total layer thickness undoubtedly have considerable influence. Suffice it to say that each photo-cathode type has its own peculiar spectral sensitivity distribution and even within one type some variation exists due to manufacturing variations.

Figure 1.2 shows the approximate curves for several common photo-cathodes and indicates their relative sensitivities in the visible region of the spectrum.

A theory, in conformity with these facts, was proposed by Einstein and our first mathematical expression for frequency conversion arises directly from his hypothesis which, briefly, is developed as follows[8].

It has been described earlier how light is absorbed to give rise to photo-electrons. Planck postulated that the light energy was absorbed and converted in discrete units named quanta or photons. Each photon had an associated energy of $h\nu$ where $h$ is Planck's constant of conversion, and $\nu$ is the frequency of the radiation.

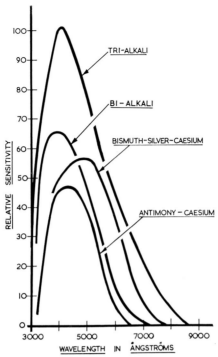

Fig. 1.2. Spectral response curves of photo-emissive surfaces.

Einstein claimed then that if a photo-electron was to be emitted by the absorbtion of a photon, there must be a strict quantitative energy conservation according to the simplified equation:

$$h\nu = eV, \tag{1.1}$$

$e$ is the electronic charge in coulombs; $V$ is the potential in volts necessary to suppress the emission of the photo-electron; $h$ is Planck's constant in joule-s; and $\nu$ is the frequency in Hertz. From this equation one can obtain explanations for the observed effects mentioned above.

$h\nu$ is linearly related to frequency and so Einstein's equation indicates that with increasing frequency, or decreasing wavelength, the photo-electron energy increases since the voltage required to suppress the emission increases.

Inevitably, some of the energy of the emitted photo-electron is expended in overcoming the surface binding forces and these will vary with the emitting material. Hence, the measured photo-electron energy varies accordingly. The energy loss is usually expressed in electron-volts and is given by $\phi$, when equation (1.1) becomes:

$$\tfrac{1}{2}mv_0^2 = h\nu - e\phi, \tag{1.2}$$

45

where $m$ is the mass of the electron in kilogrammes, $v_0$ is its maximum speed of emission in metres per second, and $\phi$ is the contact difference of potential or work function.

If the energy of the electron is measured by the voltage necessary to suppress all emission $(V)$, the equation becomes:

$$eV = h\nu - e\phi \tag{1.3}$$

or

$$V = \frac{hc}{e\lambda} - \phi,$$

which is

$$V = \frac{12\cdot4 \times 10^3}{\lambda} - \phi, \tag{1.4}$$

where $V$ and $\phi$ are measured in volts and $\lambda$ in Ångströms. Equation (1.4) shows how the photo-electron emission is dependent upon the wavelength of the stimulating light and the work function of the emitting surface.

If $V$ is zero or only just positive—i.e. photo-emission is at its threshold point, equation (1.4) defines the threshold wavelength. Furthermore, one can deduce that maximum value of work function which will allow photo-emission at a given wavelength.

Einstein's equation clearly supports all the experimental evidence discussed above and forms the basis for most mathematical treatments of photo-electricity. One practical consequence of the equation is its application to the measure of efficiency for photo-emitting layers. If one photon gives rise to one photo-electron then the surface is said to have a quantum yield or quantum efficiency of 100%. In practice this is never realized.

A further interpretation of Einstein's equation is that it defines the energy value of an electron capable of being produced by light of a certain frequency and, since we are discussing frequency conversion, one may ask 'What frequency can be ascribed to an electron?' To answer this question resort must be made to the theory of wave mechanics and in particular to the fundamental wave equation of Schrödinger[9] but these treatments are of academic interest only.

Suffice it to say that electrons may be shown to behave as waves in so far as they can be diffracted as can light. If a narrow monochromatic electron beam is sent through a very thin metal foil, a suitable electron detector will indicate a diffraction pattern similar to those obtained with X-rays or light.

For all practical purposes the emission of the photo-electrons occurs immediately the photons reach the photo-cathode. In this instantaneity lies the main difference between photo-emission and photo-conductivity.

Modern photo-conductive cells are based on impurity type materials and are produced by closely guarded recipes.

The creation of the free electrons in photo-conductors is probably instantaneous, as has been described earlier, but their arrival at the terminating electrodes of the cell, under the influence of the applied electric field, is somewhat prolonged compared with photo-emitted electrons. Due, in some part, to loss by the increased collision probability within the solid state and otherwise to the increased material thicknesses involved, this delay of detection introduces time lags of variable magnitudes in photo-conductive cells.

The thicker layers of these cells also influence their spectral sensitivity. The layer colour and its optical absorption have a selective filtering effect on the incident light.

Some representative spectral response curves are shown in fig. 1.3.

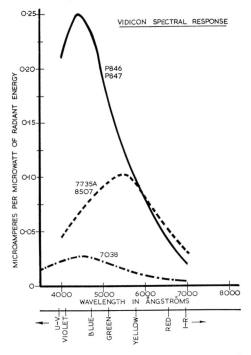

Fig. 1.3. Spectral response curves of photo-conductive surfaces as present in some vidicon television camera tubes.

PRACTICAL photo-emitting surfaces can only exist *in vacuo* or in a low pressure of an inert gas such as argon. Some kind of airtight enclosure is therefore necessary to contain such a photo-cell and the name photo-tube has been adopted for this category of device.

A simple photo-tube consists of a photo-sensitive cathode either opaque or transparent, a window to permit the entrance of light and an anode to collect the photo-electrons. At least two conducting wires through the walls of the envelope are provided so that the necessary potentials can be applied. For manufacturing reasons further electrical leads are necessary.

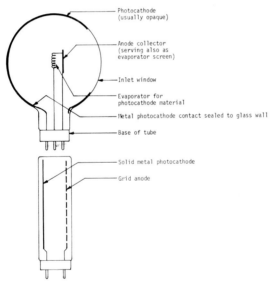

Fig. 2.1.    Diagrams of two simple photo-tubes.

Figure 2.1 shows diagrammatically two forms of such cells while fig. 2.2 shows a photograph of some commercially available photo-tubes.

The first photo-tubes were made using the metal sodium as an emitter[10] and it is interesting to note the method of producing a layer of free sodium within the vacuum. The evacuated bulb was immersed in a bath of molten sodium when, after some minutes a film of the metal

[*Courtesy of Radio Corporation of America*]

Fig. 2.2.   Photographs of simple photo-tubes.

was produced internally by straight diffusion through the soda glass envelope.   Modern techniques invariably are based on evaporation, although sublimation of chemically generated materials is often used.

For the photo-cathode some typical mixtures are shown in Table 2.1 and the choice is often made considering requirements such as spectral and overall sensitivity and process cost.

Highest sensitivity photo-cathodes of any type are best prepared in their final vacuum—i.e. without any intermediate exposure to air. Recent developments have shown that an auxiliary vacuum tube can be

| Basic elements | JEDEC* designation | Approximate spectral range† | Typical sensitivity† $\mu$A/lm | Substrate‡ |
|---|---|---|---|---|
| Ag–Cs–O | S.1 | 0·4–0·11 | 20 | SM or T |
| Ag–Rb–O | S.3 | 0·3–0·87 | 10 | SM |
| Sb–Cs | S.5 | 0·2–0·65 | 40 | SM |
| Sb–Cs | S.9 | 0·32–0·65 | 30 | T |
| Bi–Ag–O–Cs | S.10 | 0·35–0·77 | 40 | T |
| Sb–Cs–O | S.11 | 0·3–0·67 | 70 | T |
| Sb–Na–K–Cs | S.20 | 0·3–0·85 | 150 | T |
| Sb–K–Cs | — | 0·3–0·63 | 65 | T |

\* Joint Electron Device Engineering Council.
† Using 2854°K white light.
‡ SM, solid metal; T, translucent.

Table 2.1.   Some typical photo-emissive layers.

50

used for photo-cathode production which can then be relocated by a mechanical transfer process[11]. This type of tube will be considered later in the book.

For simple cells the requirement of *in situ* processing is met by mounting the basic materials in solid and inactive form inside the envelope. The activating chemicals—usually caesium—can be similarly mounted or contained in an appendix. Figure 2.3 gives sections through

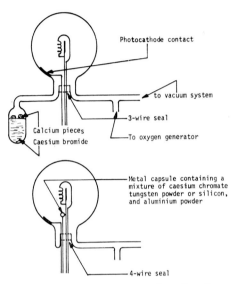

Fig. 2.3.   Methods for generating caesium for photo-activation.

typical photo-cells and shows the various components for caesium generation.

For working in the visible and infra-red spectrum there is little preference for envelope material. Hard or soft glass can be used, the main consideration being the ease of fabrication. In this area a choice between the poor thermal resistance but excellent metal sealing properties of soft soda glass is matched against the ease with which harder glasses can be worked and the increased difficulty of sealing in the metal leads. The newly developed halogen-free hard glasses are especially suitable for photo-cells and minimize sealing problems.

For photo-cells sensitive in the ultra-violet, the low wavelength cut-off of conventional glass prevents its use and a complete quartz envelope or, alternatively, a quartz window has to be used.

It is useful to discuss briefly the practical details for manufacturing the surfaces mentioned in Table 2.1. They will be taken roughly in order of discovery.

## 1. *Silver–oxygen–caesium*

The surface has the JEDEC designation S.1. The base material is silver and the base layer is conventionally either pure silver oxidized or silver oxide with an additional layer of silver. It can be opaque or transparent.

For opaque layers the silver may be chemically deposited on the inner envelope walls, be in the form of a plate, a plated base metal or produced by internal evaporation. For transparent films only evaporation is conventionally used.

The oxygen is usually generated external to the tube by heating potassium chlorate, silver oxide powder, barium oxide or potassium permanganate.

Caesium may be evolved by applying heat to a suitably contained intimate mixture of caesium chromate, aluminium powder and tungsten or silicon powder. Alternatively a side tube containing caesium bromide and calcium chips is heated and the caesium metal produced is driven into the photo-cell envelope by chasing it along the connecting tube with a hand torch.

The photo-activation process comes at the end of a normal vacuum outgassing schedule for a glass vacuum tube. Briefly, this consists of the following procedure.

The tube is sealed either by glass blowing or via a compression seal, usually of neoprene, on to the high vacuum system. After ensuring that the tube is vacuum tight by a Tesla probe, it is baked at as high a temperature as its components or glass envelope will allow. For soda glass 350°C is a fairly safe temperature while borosilicate glasses will stand up to 500°C. At some time during the bake, liquid air is applied to the freezing trap of the exhaust system. Practice differs as to its time of application but in general, when the vacuum pressure has fallen to about $10^{-3}$ torr, the freeze trap may be filled with liquid air. The severe cooling serves two purposes—firstly it traps any condensable vapours produced by the high temperature bake of the photo-cell—thereby improving the vacuum. Secondly, it prevents the ingress of vapours from the diffusion pump fluids be they oil or mercury. Some experts claim that only mercury diffusion pumps can make effective photo-cells. Pumping speed is all important and many doubt the use of vac–ion pumps in this respect. The vacuum bake of the tube is continued until a pressure of better than $10^{-6}$ torr is registered. During it any appendages, such as caesium tubes, must be heated to degas them.

With the tube cool, all internal metal is now degassed either by direct or radiofrequency heating and the tube is baked for a further short period.

From this point on the process varies according to the type of cathode to be made. For a solid metal whether it is on the tube wall or separately mounted, a clean up by glow discharge is used. If the base layer is to be prepared *in vacuo*, it is now produced either opaque or ransparent. For the solid or opaque films a high frequency oscillator is then used to produce a glow discharge with the generated oxygen. The oxidation is continued until the silver has assumed a bluish purple colour. The transparent layers, usually on the tube wall, can be either thin[12]—of a yellow colour—or thick of a blue colour. In either case the oxidation causes the layer to become colourless. For the former thin layer a second silver layer is now evaporated to a blue colour. Assistance in these evaporations may be obtained by using an auxiliary transmission measuring equipment to monitor the amounts deposited. The cell is now ready for the addition of caesium. For this, it is necessary to set up the circuit indicated in fig. 2.4.

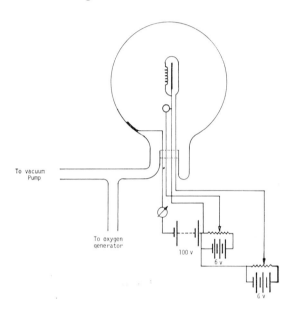

Fig. 2.4. Photo-activation circuit.

The reaction with caesium is best carried out at a temperature of about 165 °C and so the main baking oven is again applied after oxidation. A source of light is included inside the oven. At about 165 °C caesium is carefully introduced into the tube and the photo-current observed by repeated switchings of the light source. A maximum will be reached and the caesium reaction is stopped and the bake continued until a second maximum peak of sensitivity is attained. The cell is then carefully cooled without upsetting the caesium balance in the tube. Some workers try the effect of a further small amount of silver at this point—it can enhance the sensitivity considerably. Further baking can also be applied.

Opinions differ as to whether the cell should be sealed from the vacuum system hot or cold. The correct process is that which, by experiment, gives the best result.

The silver based photo-cathode has high thermal emission and this must be remembered during the photo-activation. It will be independent of illumination and can hence easily be separated from the photo-currents by a simple subtraction. Any ohmic leakage produced by the caesium vapour can also be separated from thermal or photo-emission by reversing the polarity of the saturating potential. That part of the meter current which also reverses is, of course, due to ohmic leakage. The activation process according to Sommer[13] is represented by the equation:

$$Ag_2O + 2Cs = Cs_2O + 2Ag. \qquad (2.1)$$

The photo-electrons probably originate in the caesium oxide following the argument in Chapter 1. In fact pure caesium oxide is photo-sensitive but to a much smaller extent and general opinion attributes the increased sensitivity of the silver mixture to its higher conductivity  The further enhancement of sensitivity when more silver is added at the end of the process adds credence to this theory

Applications of the caesium oxide silver photo-cell are limited due to its thermal emission  A typical relationship between thermal current and temperature is given in fig. 2.5. However, the layer is unique in having sensitivity in the long wavelength or near infra-red region and with this property it has special uses.

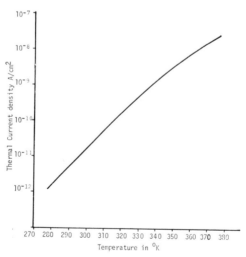

Fig. 2.5.   Relationship between thermal emission and temperature for a Silver Oxygen caesium photo-cell.

| JEDEC number | Spectral use | Usual base material | Typical sensitivity in $\mu$A/lm of white light |
|---|---|---|---|
| S.4 | Visible | Opaque metallic | 35–45 |
| S.5 | U.V. and visible | Opaque metallic | 35–45† |
| S.9 | Visible | Entrance window | 30–35 |
| S.11 | Visible | Entrance window | 50–60‡ |
| S.13 | U.V. and visible | Entrance window of fused silica | 50–60 |
| S.17 | Visible | Opaque reflecting metallic | 130 |
| S.19 | U.V. and visible | Opaque metallic | 50–60† |
| S.21 | U.V. and visible | Entrance window | 35–45 |

† S.5 has an entrance window of U.V. transmitting glass while S.19 uses used silica.
‡ Oxidized.

Table 2.2. Typical antimony–caesium photo-surfaces.

## 2. *Antimony–caesium*

This surface has the JEDEC designations S.4, S.5, S.9, S.11, S.13, S.17, S.19, S.21, dependent on its method of incorporation into the photo-tube. See Table 2.2.

It is probably the most simple photo-layer to prepare. The process consists of vaporizing caesium on to an evaporated layer of antimony in a similar manner as that by which the silver oxide is processed.

The sensitivity of the antimony–caesium surface is considerably higher than that of the S.1 but for a somewhat reduced spectral band-width. Its quantum efficiency is approximately 20%, being some fifty times that of the most sensitive silver–oxygen–caesium surface.

The specific resistance of the Sb–Cs surface is high and it can be considered almost as a true semiconductor. Its high sensitivity may be attributed to this property.

In chemical composition the Sb–Cs layer is probably $SbCs_3$. It is a pure chemical compound not an alloy or a mixture and as such lends itself more readily to analytical work on the basic theory of photo-emission and the reader is referred to the many excellent papers by Sommer and his co-workers on the subject[14].

As with the Ag–O–Cs layer the Sb–Cs can be treated with oxygen after its optimum initial processing. The equation representing the reaction is:

$$4SbCs_3 + 3O_2 = 6Cs_2O + 4Sb. \tag{2.2}$$

This expression indicates a set of conditions similar to those expressed by equation (2.1) namely a combination of caesium oxide with an interspersed metal.

Sommer postulates that the oxygenized mixture will have a lower sensitivity due to (a) its lower specific resistance caused by the free antimony; (b) the conversion of the photo-sensitive Sb–Cs to non-photo-sensitive Sb; and a higher sensitivity due to a lower work function resulting from the creation of $Cs_2O$ with its inevitable absorbed layer of free Cs.

The lower work function should also increase its red sensitivity. Each of these possibilities has been experimentally verified.

High practical importance is attached to the somewhat higher natural work function of the Sb–Cs photo-cathode. Its sensitivity to the longer wavelengths is extremely low and its thermal emission at normal temperatures is practically zero. As a consequence the Sb–Cs photo-cathode finds extensive application in photo-multiplier tubes (see Chapter 6).

## 3. *Bismuth–silver–oxygen–caesium*

This surface has the JEDEC designation S.10. It is probably the most important of the presently known photo-emitting materials but from its increased number of constituents it is obviously more complicated than the two previously described. Its importance lies in its spectral sensitivity. This can be made to approach very closely that of the average human eye—i.e. to the photopic curve. Such a photo-cathode will yield photo-currents proportional to the electric currents which are said to constitute the stimuli which the human brain receives from the eye. Unfortunately, in the photo-cell the stimuli are monochromatic, but for monochrome television purposes for instance it does enable colour scenes to be reproduced in aesthetically correct assemblies of monotones.

Classically this photo-cathode is prepared as a semi-transparent layer by applying its constituent elements in sequence to a glass base. Almost any sequence may be adopted provided of course a firm base layer is first established. Highest sensitivities are obtained if caesium is added during the later steps in the process. No success can of course derive from adding caesium or oxygen as a first operation on the glass substrate.

By some preliminary experimenting it is possible, in commercial S.10 photo-cell production, to prepare evaporators carrying a composite load of bismuth and silver mixed in the correct proportional amounts for he tube geometry involved.  A single heating cycle will then produce a layer of optimum thickness and composition requiring only oxidation and caesium activation for the best sensitivity.

The conventional oxygen glow discharge has little effect on the bismuth at normal temperatures and the intermediate layer composition is a mixture of bismuth and silver oxide, viz. Bi–Ag$_2$O and it is this layer which is reacted with caesium.  The usual caesiation temperature of about 160°C is used.  On heating, a colour change occurs and the layer becomes colourless—due, it is believed, to the reaction:

$$2Bi + 3Ag_2O \rightarrow Bi_2O_3 + 6Ag.$$

Subsequent exposure to caesium produces, from the bismuth oxide, a mixture of caesium oxide and caesium bismide.  The final photo-emitting surface is probably of the form:

$$Cs_2O + Cs_3Bi + Ag.$$

A detailed study of the photo-emission from this material is outside the scope of this present work but it must be pointed out that the role of the silver is probably different from that in the Ag–O–Cs layer. Experiment has shown that it can be replaced by metals such as gold, copper or palladium of which the only common property is their relative stability against oxygen.  Metals such as aluminium and manganese can only serve as alternatives to silver if applied after the oxygen has been fixed.  It may be deduced that silver plays no specific part in contributing to photo-sensitivity and also that its function is not merely to increase the electrical conductivity.  Its most likely function according to Sommer and Spicer[15] is to increase the depth of the layer from which photo-electrons can be produced.

## 4. Multi-alkali

The term multi-alkali is synonymous with tri-alkali and the two conventionally refer to that surface which has the JEDEC designation S.20, and is composed of antimony, potassium, sodium and caesium[16].

As the quantity of its constituents suggests, it is an extremely complicated material with an involved manufacturing schedule but yet conforming to a basic simple pattern.  It is one of the family of alkali antimonides of which antimony–caesium, described earlier, is the simplest.

The discovery of the multi-alkali photo-cathode makes interesting reading[17].  In trying to repeat earlier experiments aimed at combining antimony and lithium, the heaviest alkali metal, Sommer decided to analyse his evaporated layer and found no traces of lithium but only potassium and sodium.  This was explained after further analyses by finding that all commercially pure lithium salts contain traces of potassium and sodium.  It was these traces, of metals of lower melting points than lithium, which were sufficient for the activation of the photo-layer. Subsequent work did succeed in producing antimony-lithium cells[18].

It may be said in generalization that the formula $SbM_3$ represents the family of alkali antimonides. M can be Cs, K, Na, Rb, Li, or any combination of them, provided the valency rules are satisfied. For example, the material $SbNa_2K$ is a good photo-emitter. Certain combinations of the above elements are more sensitive and it is on these and on Sb–K–Na–Cs in particular[19], that emphasis will be confined.

The empirically established basic composition of the material is $SbNa_2K$ and this is unchanged by the addition of Cs although the work function is lowered—so enhancing the sensitivity. Sommer writes the complete formula $(Cs)SbNa_2K$, indicating a caesiated sodium potassium antimonide.

The preparation of the S.20 surface is probably the most complicated photo-activation process.

The general photo-activation temperature is similar to that for the previously described photo-cathodes and the first step is to obtain the highest photo-sensitivity with potassium alone—this is only about $\frac{1}{30}$ $\mu A/lm$. Step two is to evaporate antimony on to the cooled base until an opacity of about 75–80% is reached. Then, with the cell at photo-activation temperature, further potassium is released until a higher peak of sensitivity is indicated—about 5 $\mu A/lm$. Invariably, the potassium control fails to respond sufficiently quickly and the peak is passed. However, sodium is now released with the cell at a slightly higher temperature and the peak is rapidly regained and usually enhanced, but the sodium treatment is continued until about 80% of the new sensitivity peak is indicated. The conventional temperature having been re-established, more antimony is evaporated until a further fall in sensitivity of some 70% is established. A further treatment of potassium rapidly restores the sensitivity and this is continually improved by successively evaporating small amounts of antimony and potassium until a sensitivity figure in the region of 50 $\mu A/lm$ is attained. At this stage caesium is released and a new peak up to four times the previous one is obtained. Again, an alternation of antimony and caesium will enhance this sensitivity which will rise as the cell cools to a figure often in excess of 200 $\mu A/lm$.

Variants on the above process are common—each worker being somewhat of an artist in his recipe for obtaining maximum sensitivity.

The tri-alkali layer has one special process peculiarity—its dislike for oxygen. Oxidation invariably destroys or seriously reduces its photo-sensitivity.

The very high sensitivity of the tri-alkali layer makes it ideal for photo-cell and photo-multiplier application. Its non-panchromatic spectral sensitivity makes it of doubtful use in television pick-up tubes. This topic will be discussed in a later chapter.

## 5. *Bi-alkali*

At the time of writing this photo-cathode has no JEDEC designation. This material may be described as a simplified multi-alkali surface since, not only does it not contain sodium but its preparation is much easier.

As with the S.20 surface, firstly the glass substrate is photo-activated with potassium. Then, after the short higher temperature bake, antimony is evaporated to a transmission of about 75%, further potassium is added to peak sensitivity and finally caesium is vaporized to produce the final rise at the activating temperature of 165°c. After cooling comes the main departure from the multi-alkali process—that of oxidation. Oxygen is admitted to a pressure of about $10^{-4}$ torr until the sensitivity ceases to rise.

In application, the use of the bi-alkali photo-cathode is somewhat restricted due to its higher specific resistance. For photo-multipliers, where the input light level is very low, it is ideal. It is of limited use at higher levels and the operating level of some television camera tubes is near the upper limit.

If the input light level exceeds this limit the large currents produce a voltage change across the layer to cause a departure from linearity in a simple photo-cell and a defocusing in an image tube.

One significant feature of the bi-alkali photo-cathode is its very low thermal dark current—a distinct advantage in photo-multipliers and image tubes working at low light levels.

In structure the bi-alkali photo-cathode conforms to the general alkali antimonide composition $SbM_3$ and recent work suggests[20] as its basic formula $K_2CsSb$ with oxidation having the same effect as with $Cs_3Sb$ producing a red enhancing micro layer of $Cs_2O$.

## 6. *Gas filled photo-tubes*

If a simple photo-tube contains an atmosphere of an inert gas such as argon and the cathode/anode potential is above the ionization potential of the gas then amplification of the photo-cathode current occurs. A common pressure of argon is 0·1 torr Hg and amplification factors of ten are usual. Amplification occurs also from the release of secondary electrons at the photo-cathode as a result of positive ion collisions—the positive ions being produced by the ionization process. Detailed discussion of gas filled tubes is outside the scope of the present volume.

# CHAPTER 2.3

## photo-conductive cells

IT was in 1873 that telegraph engineers May and Willoughby-Smith noticed that variations in sunlight intensity produced sympathetic changes in the values of their selenium resistance boxes[2]. This was the discovery of the photo-conductive effect.

Photo-conductive cells are manufactured from materials possessing the properties outlined in previous chapters and are basically insulators. To sensitize them they are processed to include some impurity or imperfection which causes the material to possess an intermediate energy level between the filled or valence band and the conduction band. Conduction can then take place in one of two ways. Firstly, when electrons are excited from the impurity level to the conduction band—these are termed n type or donor materials. Secondly, electrons can be excited from the filled band to the impurity level to leave holes in the former—these are p type or acceptor materials. In either case the energy gap is small enough such that photons have sufficient exciting energy.

Materials such as cadmium sulphide and selenide, indium antimonide are typical examples of impurity or imperfection photo-conductors and are used in commercial photo-cells.

A second type of photo-conductive photo-cell depends upon a barrier or junction system. An n-type material in intimate contact with one of p type develops across the contact zone an electromotive force which can either, as described earlier, be detected directly in an outside circuit or, if biased in the reverse direction by an external circuit, behave as a photo-conductor. Such systems, of which germanium is a typical example, are called junction or barrier cells. A photo-transistor is a particular type of a junction photo-cell.

## photo-voltaic cells

THE photo-voltaic cell is probably the simplest for every day use, needing only a voltage or current detector. In general, its main use is for detection of light rather than its measurement, although many brightness measuring equipments utilize a photo-voltaic cell.

The modern photo-voltaic cell is exclusively a junction photo-cell in solid materials. Various materials are capable of being processed into both the p-type and n-type forms whence a composite sandwich can be formed into a pn junction. One can illustrate the situation at the junction as in fig. 4.1.

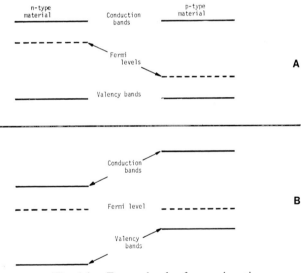

Fig. 4.1.   Energy levels of a p–n junction.

As indicated for the n-type material the Fermi level lies near to the conduction band while in the p type it lies near to the valence band. When in contact an electron/hole interchange takes place until a common Fermi level is established across the junction to produce a contact potential between the two materials. The p-type material becomes negative with respect to its n-type neighbour. When the junction is illuminated electron/hole pairs are created and flow across it to produce a current in an external circuit. Typical materials for the manufacture of photo-voltaic cells include the following: silicon, germanium, selenium, indium antimonide, indium arsenide and gallium arsenide. Commercial manufacturing processes are not available for publication and only a general outline of a few of them can be given.

## *Selenium* (courtesy of Evans Electroselenium)

The base of the cell is a steel plate and to it is applied molten selenium. After a pressure/heat treatment a thin transparent metal layer is deposited by cathodic sputtering and to this is applied a contact strip of a low melting point metal. The whole is then protected by a lacquer spray. In use the steel backing plate becomes the positive and the soft metal strip the negative contact. The barrier layer is the interface between the selenium and its transparent metal coating.

## *Silicon* (after RCA)[21]

The cell base is a thin single crystal slice of p-type silicon and on to it is formed by diffusion a layer of n-type material about 0·5 micron thick. For a contact to the base silicon a continuous layer of soft solder is used while a metal grid is applied to the upper surface of the thin n-type layer. Silicon cells have been developed to a high degree of efficiency as solar cells for use in space satellites, etc.

## secondary electron emission

THE inclusion of this chapter is a necessary convenience as an adjunct to succeeding chapters. It will deal with the production of electrons by electrons as distinct from photons which have been the interest so far.

Soon after the introduction of the thermionic valve it was realized that, under certain conditions, electrons bombarding a surface would cause it to emit further electrons. It was shown that this production of secondary electrons—the phenomenon of secondary emission—was equally dependent on the energy of the incident, or primary electrons and the nature of the surface, as was photo-emission.

An energy/emission curve similar to a spectral response curve for a photo-emitter could be drawn. In this case one deals with electron currents only and a simple vacuum tube circuit is necessary for the determination of the curve.

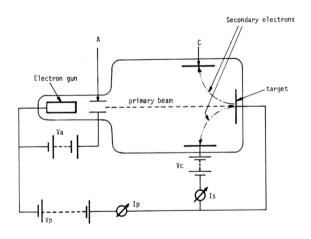

Fig. 5.1.   Tube and circuit for measuring secondary emission.

In the diagram (fig. 5.1.) is represented a vacuum tube containing a conventional electron gun, its final anode $A$, a target, and an auxiliary, or second anode $C$. The gun anode is held at a fixed potential $V_p$, while the potential of the target can be varied by $V_p$ which is adjustable from zero to say 2000 volts. Between the target and the cylindrical electrode $C$ is maintained a fixed low potential of say 100 volts.

With the thermionic gun energized and the other potentials applied, currents will be indicated in the two meters. That of the primary electron beam will be measured as $I_p$ while secondary electrons produced at the target will be attracted by $V_c$ to the second anode and be registered as $I_s$ which is the secondary electron current.

The secondary emission ratio or coefficient (SEC) is taken as $I_s/I_p$ and if it is plotted as a function of $V_p$, a curve, similar to that shown in fig. 5.2, is obtained.

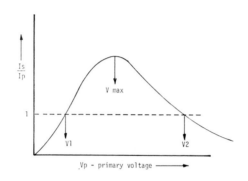

Fig. 5.2. Relationship between secondary emission and primary electron energy.

No scales are shown in fig. 5.2 in order to make it more general. Different materials will have different values of $I_s/I_p$ (peak); $V_1$, $V_p$ and $V_2$. Table 5.1 shows typical values of these parameters for several materials. For the investigation of insulators on the target a pulse technique using ballistic galvanometers is used.

The general shape of fig. 5.2 is easily understood from a purely logical argument. Assuming that secondary emission is purely a collision process, it may be expected that as the primary electron energy increases so does its penetration and the probability of collision with the atoms of the material increases. Further penetration due to higher voltages will cause more secondary electron *production* but, due to the greater depth of penetration these electrons, which are of low energy, have greater difficulty in escaping to the vacuum and hence the secondary electron *emission* decreases.

The phenomenon of secondary electron emission is of considerable importance in electro-optical devices. As will be made evident later, its

| Material | δ (max) | V peak |
|---|---|---|
| Silver | 1·5 | 800 |
| Gold | 1·46 | 750 |
| Bismuth | 1·15 | 550 |
| Beryllium | 0·53 | 200 |
| Boron | 1·2 | 150 |
| Caesium | 0·72 | 400 |
| Potassium | 0·75 | 200 |
| Magnesium | 0.95 | 300 |
| Nickel | 1·3 | 350 |
| Titanium | 0·9 | 280 |
| Black nickel†† | 0.45 | 700 |
| Magnesium silver (2% Mg)† | 8·0 | 600 |
| Potassium chloride | 7·5 | 1000 |
| Caesium antimonide | Up to 8·3 | Approx. 400 |
| Beryllium oxide | 3·4 | 2000 |
| Magnesium oxide | 4·0 | 400 |
| Aluminium oxide | Up to 4·8 | Approx. 1250 |
| Mica | 2·4 | 380 |

†† Nickel evaporated in an atmosphere of a rare gas.

Table 5.1. Secondary emission characteristics of some typical photo-tube materials. (Extracted from Bruining, *Physics and Applications of Secondary Electron Emission*, Pergamon Press 1954; except for † which was taken from Miyashiro and Hirashima, *J. Phys. Soc. Japan*, **12,** 771, 1957.)

presence can either hinder or facilitate the achievement of certain design requirements. In this present chapter it is convenient to discuss several states of the target in a tube having the configuration of fig. 5.1. In the diagram it is shown held at $V_p$, but if it is insulated the incident beam, $I_p$, can be arranged to strike it by inertia, so to speak. Then according to the beam energy one of several possibilities can arise.

If the beam energy is less than that represented by $V_1$ in fig. 5.2 then the net electron input to the target will be positive—its potential will fall continuously until it becomes just so negative that the primary beam cannot land. At this point the target is fractionally below the potential of the thermionic gun and it is said to be cathode potential stabilized.

If the primary beam energy corresponds to a potential above $V_1$ and below $V_2$ then the target tends to lose electrons and becomes more positive. It will rise in potential as long as the secondary electrons can be collected, i.e. while a positive field gradient exists in front of it due to the presence of an electrode at a higher positive potential. In fig. 5.2 this is the cylindrical anode collector $C$. The target stabilizes at this, the so-called second anode, potential.

At beam potentials above $V_2$, by similar reasoning, it is seen that the stable potential of the target is $V_2$. Unstable target and tube conditions arise if beam energy corresponds exactly to $V_1$ or $V_2$.

Materials differ widely in their secondary emission coefficients. Table 5.1 gives the values for a few of those commonly used in vacuum tubes together with the approximate potentials for their peak values.

The law of the conservation of energy is applicable to the phenomenon of secondary electron emission inasmuch that the larger the number of secondary electrons the smaller are their individual energies in relation to that of the incident primaries.

The secondary emission properties of insulators is of particular interest in light conversion devices and recent work[22] has shown that an internal effect opens up the possibility of extremely high sensitivities. It has been discovered that insulators of high porosity, i.e. of an open fibrous nature, can be made such that secondary electrons released within the material will make further internal collisions before reaching the main material/vacuum boundary. At each collision more secondaries are produced and an avalanche effect ensues. The final electron shower can be either collected at the surface of the porous film or caused to be incident upon a further film to achieve added amplification.

The phenomenon is known as secondary electron conduction and is colloquially referred to by its initials SEC. An alternative term is transmission secondary electron multiplication or TSEM. Conversion tubes incorporating materials with these properties will be described in later chapters.

## photo-multipliers

A CATEGORY of electron tubes which combines with a simple photo-emissive cell an electron multiplier of the secondary emission type is popularly, but many claim wrongly, termed a photo-multiplier. The tubes are not, as the name suggests, light amplifiers—this latter type of tube will be described in a later chapter. The name ' multiplier photo-tube ' is more correct but less acceptable.

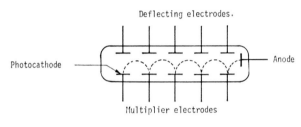

Deflecting electrodes.

Photocathode

Anode

Multiplier electrodes

(Magnetic field perpendicular to diagram)

Fig. 6.1 (a) Electromagnetic photomultiplier.

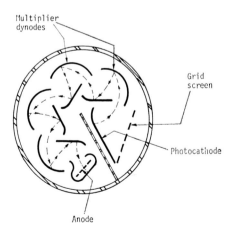

Multiplier dynodes

Grid screen

Photocathode

Anode

(b) RCA 931-A type photomultiplier.

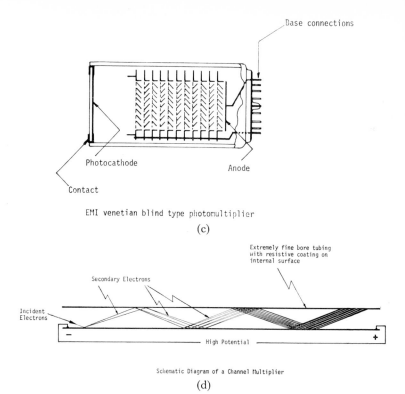

EMI venetian blind type photomultiplier

(c)

Schematic Diagram of a Channel Multiplier

(d)

Fig. 6.1. Diagrammatic versions of four electron multipliers.

A photo-multiplier is an extremely sensitive photo-cell due to the inclusion within the same envelope of a multiplier section consisting of one or more multiplying electrodes or dynodes into which the photo-electrons are directed by appropriate electron lenses before being collected as output current. Whereas, in practice the maximum sensitivity obtainable with a photo-cell is about $250 \, \mu\text{A/lm}$—photo-multipliers are built with sensitivities up to $1000 \, \text{A/lm}$. The light

Schematic Diagram of Schaetti type
Electron-Multiplier

Fig. 6.2. Schaetti photomultiplier.

detectability of such a tube becomes of the order of $10^{-20}$ lm—dependent on the current measuring equipment. Single photons can thus be detected. The application of such devices to astronomy and scintillation detection is immediately obvious.

The possible combinations of a photo-cathode with the many available types of dynode are almost infinite. Figure 6.1 a, b, c, d, shows four of the most frequently used designs.

Early photo-multipliers were of type 'a', where the dynodes were mounted ladder-rung fashion in a long envelope. The inter-dynode electron paths were controlled by a magnetic field combined with stepped potential supplies.

Fig. 6.3.    Photograph of 931-A photo-multiplier.

Simplest in concept, type ' a ' serves to indicate the basic require-ments. A low impedance potential divider network is provided with tappings for each dynode such that the potential difference between successive tappings is somewhere near the optimum for the particular dynode surface used. Cooling and/or temperature control will be pro-vided as the system dictates. Fig. 6.2 shows an alternative, non-magnetic version, of the principle.

A tube such as type ' a ' is obviously inconveniently large and a condensed version was evolved. It is shown as type ' b '. In this design the dynodes are now curved and interleaving one with the other. Electron focusing between them is achieved purely electrostatically by careful electron optical design of the electrode system. Such a structure is limited in flexibility inasmuch that the number of dynodes cannot be varied without a complete re-design. A further main limitation is the inability to prepare the dynode surfaces in the final vacuum. The compact structure leaves no space for the inclusion of processing capsules, etc. A photograph of the tube is shown in Fig. 6.3.

A design aimed at improving these features is that incorporating the so-called 'venetian blind' dynode. With such a tube, primary electrons are incident upon one face of each dynode whereas the secondary electrons are, in effect, emitted from its opposite face. Type 'c' shows the arrangement and Fig. 6.4 shows a single dynode.

[*Courtesy of E.M.I.*]

Fig. 6.4. E.M.I. type venetian blind dynode.

A highly transparent metal mesh, mounted on the input dynode face and held at the same potential, prevents the retarding field of the previous stage interfering with the emission of the secondary electrons. Between each dynode is mounted a pre-coated evaporator wire which can be heated when required by a radiofrequency coil applied outside the tube. The photo-cathode can be prepared in a similar manner but caesium, for the final steps in the process, is invariably generated external to the tube since better overall control is possible.

Mesh type dynodes as used in design are not particularly efficient due to the difficulty of withdrawing the secondary electrons. For some purposes, they are mechanically and geometrically convenient.

Type 'd'—the channel multiplier has not, up to the time of writing, seen commercial acceptance but the idea augurs well—perhaps not so much for photo-multipliers but for picture multipliers—alternatively known as image intensifiers.

The basic difference between a photo-multiplier and an image intensifier introduces one extremely important property of the former. Whereas an image intensifier performs the function of increasing the brightness of a two-dimensional optical image and more often than not a static image, a photo-multiplier is asked to produce an electrically

70

recordable signal from a light input having dimensions of time and magnitude only. Rarely does a positional requirement exist.

Ideally, the output of a photo-multiplier must, then, bear an exact relationship to its input. This demands that light input *at any point* on the photo-cathode must be faithfully recorded as output and that there must be no difference in character between outputs resulting from similar inputs at various points on the photo-cathode surface. In the oft occurring condition of pulsed light inputs, their time separation must be reproduced faithfully at the tube output. Summing up these requirements it can be said that ideally the transit time of a photo-electron in a photo-multiplier must be constant and independent of position of input. It is further advantageous that the output pulse width be kept as short as possible to allow high-frequency observations.

Much of the variation in transit time derives from the electron optics of the space between the photo-cathode and the first dynode and the wide variety of shapes and electrode configurations in commercial tubes is indicative of the many attempts to optimize conditions.

A further influence on design is the desire to eliminate or, at best, minimize effects due to light produced within the tube itself. Tubes containing fifteen dynodes can need an overall potential upwards of 3000 volts. At these values, cold emission from the caesium contaminated internal structure is readily produced unless care has been exercised in component design. Similarly ohmic leakage paths across insulating supports can cause trouble. Inadequate dynode insulation will, of course, generate currents in the output stages quite dissociated from any light input and the overall signal to noise ratio of the device suffers.

Among the various designs of individual dynodes the antimony–caesium surface is extensively used as the secondary emitter. It can be prepared either by evaporating antimony on the dynode base metal before tube assembly or by including in the tube evaporators which can be energized during the subsequent vacuum processing and before the admission of caesium. Other suitable dynode surfaces not needing caesium include magnesium oxide, beryllium and beryllium copper.

Gallium phosphide activated with caesium has a secondary emission ratio of thirty at 600 volts[23] and may appear as an effective material especially for use on the first dynode where high gain is particularly important.

# CHAPTER 2.7

## television camera tubes

THIS chapter deals with vacuum tubes in which much of the substance of preceding chapters is combined to produce frequency convertors of the most intricate design. They are the picture transducers, the main, if not the only, vacuum tubes in a television camera. (N.B. This does not include cathode ray tubes used in electronic view-finders, since they are not essential features of television cameras.)

Limitations of space dictate that detailed discussion be confined to currently used tube types—probably the last vintage of vacuum tubes to be used for this purpose. Progress in solid state devices is making great strides.

However, it is felt that a brief history of the evolution of the present day tubes is appropriate in this book, even if only to justify a list of references! It is assumed that the fundamentals of a television system are fully understood [24].

Television enables a two-dimensional arrangement of various light intensities—a picture or optical image—to be instantaneously transmitted electronically between two widely separated locations. The transmitted signal originates in the television camera.

Naturally, a planar array of simple photo-cells each connected by wire in orderly fashion via a suitable amplifier system, to a corresponding array of incandescent lamps would work. Such a system has obvious limitations and to eliminate its impracticable but unavoidable multi-cable link some means of analysing the picture point by point is necessary. Then, provided that the process is rapid, the picture can be re-constituted point by point. Analysis or scanning by mechanical means was [25] proved inadequately fast and resort was made to electron beam scanning. The first published proposal of such a system was made by Campbell-Swinton in 1908 [26]. Then in the early 1930's it was indicated [27] that in the 1920's Zworykin had conceived a vacuum tube aimed at converting an optical image into a time varying electric current capable of radio transmission and subsequent re-assembly on the phosphor of a cathode ray tube. The Zworykin tube—the Iconoscope [28]—is the base from which all current designs grew. Figure 7.1 shows a picture of the tube and fig. 7.2 is a diagrammatic cross-section.

The main component of the tube is the mosaic of minute photo-cells formed on the mica support. The photo-emitter is silver–oxygen–caesium since evaporated silver has the unique property of forming itself into mutually insulated globules when subjected to heat treatment at

72

Fig. 7.1.   Photograph of iconoscope.

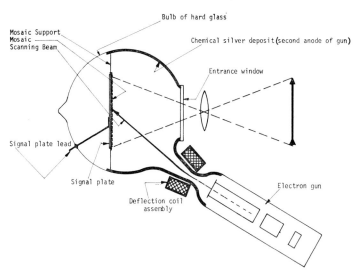

Fig. 7.2.   Diagram of iconoscope.

about 300°C. By processes described earlier, these silver islands are rendered photo-sensitive.

The tube was designed to function as follows. The optical projection of an image on to the mosaic would cause photo-emission point by point proportionate to the light intensities in the picture. Each mosaic element would consequently lose some electrons and become charged positively. When the electron beam scanned over the mosaic, each positive charge would be neutralized and a succession of pulses would be capacitively induced in the metal signal plate backing the mosaic, and hence through the signal resistor to earth. In practice this did not happen.

Since for focusing purposes a high velocity scanning beam was necessary the mosaic became stabilized at the second cross-over point (see Chapter 5), and was in fact sufficiently positive in potential that no attractive field existed for the extraction of the photo-electrons. However, the tube did function because at the point of impact of the scanning beam, high secondary emission occurred and the point was driven momentarily more positive than its surroundings. This enabled photo-electrons to leave the neighbouring areas so that, as the beam moved on, it encountered some positive charge to be neutralized and allow signal production.

The tube had low sensitivity since photo-emission could only occur for a short interval of time instead of the full television frame period as was intended. In other words full frame charge storage was not possible[29].

Since the tube design demanded that the projection of either the optical picture or the scanning beam raster be not normal to the mosaic plane, obvious geometrical design difficulties arose. The so-called keystone scanning circuit was developed to cater for the beam problem[30]. Even so, due largely to its difficult shape, the performance of the tube was far from satisfactory. Picture shading was almost unavoidable in spite of the complicated correction circuits eventually evolved.

Lubszynski and Rodda saw that a decided improvement in sensitivity could be obtained by adding a section of image intensification and in 1936 was announced the super-emitron[31] (fig. 7.3).

This tube had two important advantages over the iconoscope. Firstly, it separated the functions of photo-emitter and charge storage mosaic to allow considerably higher fundamental photo-sensitivities to be obtained. Secondly, a net gain in mosaic sensitivity derived from the high energy of the input photo-electrons resulting from the high potential at which it operated relative to the photo-cathode.

Shading problems still were a problem and Theile substantially cured them with his photo-electrically stabilized iconoscope—fondly known as the pesticon[32]. This tube was made of a smaller size than previously. In addition, strategically located pea lamps attached to its outer surface caused photo-emission to occur from its internal walls and also from the two auxiliary photo-cathodes. The resulting photo-electron spray was

Fig. 7.3.   Photograph of Super Emitron.

attracted to the higher potential edges of the mosaic and prevented their assuming values which contributed to picture shading.

At about the same time as Lubszynski announced his super-emitron he also realized that angular scanning would have to be eliminated to enable improved and more consistent control of signal production.   His proposal[33] was to cause the scanning beam to approach the mosaic orthogonally and at zero velocity over the entire raster.   He achieved this by combining the normal magnetic deflection field with an axial magnetic field in conjunction with a new electron optical lens.   Attempts to produce orthogonal zero velocity beam scanning by a combination of electrostatic and magnetic deflection had produced a tube called the orthicon[34]—it was of unacceptably low sensitivity. During the second world war came the development of a tube combining an image section, orthogonal all magnetic scanning and an additional feature, an inbuilt electron multiplier, the image orthicon[35]. Its photo-electric efficiency was not high and soon followed a similar tube, the cathode potential stabilized emitron[36], (CPS emitron) with improved photo-efficiency resulting from a photo-electric mosaic produced by novel means[37]. The CPS emitron had no image section and could be classified as an orthicon (fig. 7.4).

Fig. 7.4. Photograph of CPS Emitron. [Courtesy of E.M.I.]

The tube had considerable success but was eventually superseded by the image orthicon which in the meantime had been considerably improved in performance. The image orthicon became the most widely adopted television pick-up tube and will be described in detail.

Throughout the entire period described above covering the successful development of a photo-emissive camera tube, vacuum technologists had been attempting to utilize the photo-conductive effect for a similar purpose. This tube would depend upon local changes in the conductivity of a layer of photo-conductive material on to which an optical image was projected. A scanning beam exploring the surface would then induce in the fully conducting backing layer a current proportional to the change in conductivity. Such a tube incorporating Lubszynski's orthogonal scanning beam, which had reduced most of the difficulties encountered in the development to the point of tolerability, was announced in 1950 and named the vidicon[38].

This very simple tube gave an impetus to the spread of television for other than entertainment purposes. Its photo-conductive layer was a type of antimony trisulphide. A recent development of the tube type utilizing instead a modified lead oxide layer working as a strict semi-conductor has probably the highest commercial potential of any vacuum television camera tube so far produced.

These two photo-conductive tube types will be described in detail.

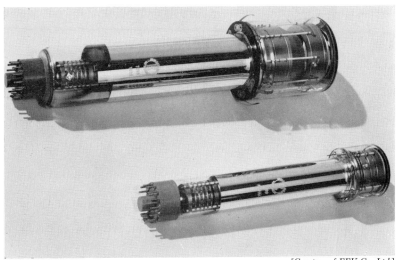

Fig. 7.5. Photographs of 3 in. and $4\frac{1}{2}$ in. image orthicon.

## 1. The image orthicon

The first image orthicon was announced in 1946 by RCA. It had a maximum diameter of 3 in. and was about 15 in. long. Its picture producing qualities were limited by its size and RCA in 1948[39] announced a larger tube of $4\frac{1}{2}$ in. diameter. This tube was not manufactured commercially in the U.S.A. but was improved in design and produced in England by English Electric Valve Company and announced in 1960[40]. Figure 7.5 shows the two tubes while in fig. 7.6 are diagrammatic sections to aid the following explanatory text.

The picture to be transmitted is projected through the optically worked window on to the photo-emissive cathode formed on its inner face. This may be either of those described in Chapter 2.2. The resulting photo-electron image is focused by means of an external axial magnetic field and internal electrostatic fields on to the target. This is a very thin lamina of a complex glass where, due to the emission of secondary electrons to the target mesh, an array of positive charges corresponding in intensity and geometrical distribution to the pattern of light and shade in the original scene, is produced. All colour informaton will, of course, be translated according to the spectral response of the photo-cathode.

Due to the relatively high capacity between the two faces of the target, resulting from its extreme thinness (about 0·0001 in.), the charge image is transferred rapidly and without noticeable diminution in magnitude, to its opposite face where it is scanned by the electron beam generated in the thermionic gun at the other end of the tube to produce the video signal.

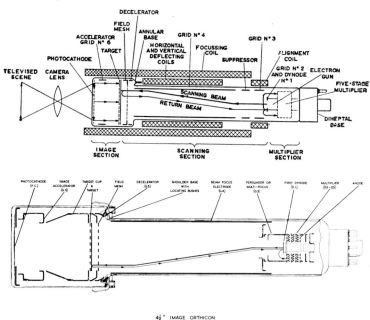

[*Courtesy of EEV Co. Ltd.*]

Fig. 7.6.   Diagrams of 3 in. and 4½ in. image orthicon.

The mechanism of scanning is as referenced earlier. The electron gun is a simple triode and has as its anode a disc containing a central aperture of about 0·0015 in. diameter. The emergent beam proceeds along the tube towards the target under the energy imparted to it by the anode potential—some 300 volts. A direction of travel parallel to the tube axis is ensured by applying a magnetic alignment field at the point of emergence. The beam is then guided by the axial magnetic field. Upon entering the region of the deflection coils (see fig. 7.6) it is deflected by the combined field of these and that of the axial solenoid by an amount determined by the instantaneous strength of the deflecting coil

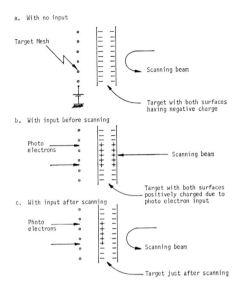

a. With no input

Target Mesh

Scanning beam

Target with both surfaces
having negative charge

b. With input before scanning

Photo
electrons

Scanning beam

Target with both surfaces
positively charged due to
photo electron input

c. With input after scanning

Photo
electrons

Scanning beam

Target just after scanning

Fig. 7.7. Target charge and discharge mechanism.

current. Upon leaving the deflection field the beam reverts to its former axially parallel direction and arrives at the target normal to the surface. If the target is uncharged—i.e. has received no photo-electrons, then scanning beam electrons will land upon it driving it negative, until it assumes a potential at which further landing is not possible. The target then becomes stabilized at a potential just below that of the thermionic cathode—it is said to be ' cathode potential stabilized '. With the target of the tube in this condition the full scanning beam is reflected by the target back to the gun (fig. 7.7).

When the photo-cathode is exposed to light to produce photo-electrons the target becomes positively charged and as the beam scans, some of its electrons will be absorbed to neutralize this charge to produce a negative charged state and cause the remainder of the beam to be reflected back to the gun.

Immediately the scanning beam passes on from a target picture element, the negative potential locally produced there conductively lowers the positive value of the input surface to a new lower level. It is from this level that more secondary electrons can leave to produce a new value of target charge according to any changed picture detail upon the photo-cathode. This cycle of potential changes and scanning beam modulation is repeated at whatever rate the television system is degined to operate.

Hence if an optical image is incident upon the photo-cathode to produce a pattern of positive charges upon the target, a beam of

Fig. 7.8.　Photograph of image orthicon gun.

continuously varying intensity is returned to the gun. It is this beam, intensity modulated by the picture charge information on the target which becomes the video output signal of the image orthicon. It must be noted that the magnitude of this signal varies inversely as the elemental picture brightness—a feature of the tube which is most disturbing as later discussion shows.

The magnitude of the modulated scanning beam is somewhat less than one-tenth of a micro amp whereas its modulation, corresponding to the input light values, is as low as 1 nA. These currents are too small for satisfactory amplification by conventional thermionic means and so an electron multiplier is included in the image orthicon. Working as a normal photo-multiplier but with no photo-electron input, the five-stage electron multiplier in an image orthicon is unique in its construction. The second, third and fourth dynodes consist of annular venetian blinds with radial slots, the first is formed on the anode of the thermionic gun and the fifth is a solid annulus. The whole is built around the electron gun as seen in fig. 7.8.

The electron path from the target to the final output terminal is a complex one. Upon arrival at the gun's anode, secondary electrons are produced with quite random emission directions but, generally, back towards the target. To guide them into the multiplier structure an electrode, variously called the persuader, or multi-focus, is positioned in front of the first dynode and is held at a slightly negative potential

80

relative to it.    Behind dynode No. 1 the second dynode, the first slatted one, is held relatively at a much higher positive potential to D.1.    The electrostatic field resulting from the potentials applied to these three electrodes causes the secondary electrons from D.1 to be constrained into an umbrella-like shower to flow into the multiplier.    The venetian blind dynodes, sometimes called pinwheels, are arranged to have successively vanes of opposite inclination to avoid possible bunching of the electron stream.    Each dynode is provided with an equipotential mesh screen in a manner similar to that described under photomultipliers. The final dynode, D.5, is solid and at it, another reversal in the direction of the electron flow occurs.    This arrangement is convenient as it ensures full interception efficiency unlike the pinwheels which, due to their method of manufacture, are only about 75% efficient.    The multiplier anode is a highly transparent mesh held sufficiently positive to D.5 to collect all the secondary electrons produced there but at a potential low enough not to materially obstruct the flow of electrons from D.4 through it to D.5.

The above represents a somewhat theoretical and idealized conception of the functional processes of an image orthicon.    However, as with all devices, and more so with those depending on a vacuum, practical considerations demand certain compromises.    It is perhaps, appropriate to devote some attention to these and to tube construction and examine the demands of the latter on the theoretical requirements and further, to analyse the interdependence of the various tube components which influence picture reproduction.

The assessment of picture quality cannot be objective.    It is almost entirely subjective and any objectivity can only be introduced as a function of majority opinion with the added proviso that this majority be sufficiently qualified to have an opinion!    The oft used phrase of the art world ' I know what I like ' applies to an infinitely greater degree to picture making by television because, unlike all other media, the television design engineer provides the observer with means to change the picture he is viewing.    It might be argued that any intelligent observer can be allowed to compare an original with its reproduction and that his (or her) opinion consequently be respected.    Unfortunately, it is an indisputable fact that the television process is incapable of producing an exact replica of the scene to which the camera is directed.    The reasons are many—the two most important are firstly, the replica is minus one dimension—two instead of three.    Secondly, the replica is either in shades of grey or artificially coloured, the original is in natural colours suitably processed by that most flexible of viewing instruments— the human eye.    The conversion of colour to its theoretical monochrome brightness equivalent can be done but the result may not be pleasing to the observer.    Fortunately, he rarely has the opportunity to view the original and its replica simultaneously!    Television colours are not the same as natural ones—the transmission system has been deliberately so chosen for practical convenience to prevent compatibility even if the reproducing phosphors allowed true colour; they do not. However, television colour is most attractive and, generally speaking, perfectly acceptable—thanks once again to the adaptability of the human eye!

To add to the confusion, the vagaries of the camera tube prevent its design engineer from stating a ' yes–no ' or ' go–no–go ' technique for operating it.

Such philosophical difficulties are minimized if one uses as a ' picture ' some form of test chart[41]. The ' test chart ' must of itself be controversial in so far as its tone scale, contrast range, information dimensions and arrangement, black to white area ratio, etc., are concerned. For this particular discussion the latter difficulty will be avoided by not specifying any particular test chart but merely using an anonymous one as a comparator for its reproduction. It will contain only white, black and shades of grey. Figure 7.9 shows some typical charts.

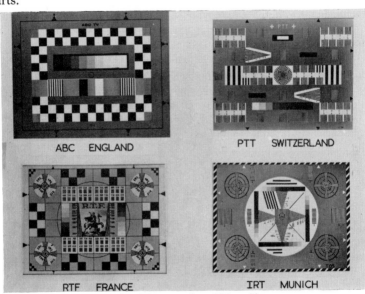

Fig. 7.9. Photographs of Test Charts.

From the image of the test chart on the cathode ray tube screen the following performance parameters of the system can be assessed:

(a) Geometry.
(b) Resolution.
(c) Gamma.
(d) Ratio of signal to noise.
(e) Sensitivity.
(f) Picture uniformity.
(g) General ease of tube operation.

Variable potential and magnetic field controls are provided on the camera control unit (CCU) to optimize picture quality (fig. 7.10). Other electrostatic and magnetic fields are fixed or pre-set as a result of operational experience. Due to the number and interdependence of the variable controls it is convenient to adopt a fairly rigid sequence of control manipulation when setting a camera in operation—such a typical recommendation is outlined below. The significance of the various terms used will be apparent as the chapter proceeds.

[*Courtesy of The Marconi Company Limited*]
Fig. 7.10. Photograph of CCU panel.

Before the camera is switched on it is preferable to allow a small amount of light to enter the taking lens by setting it to a very small aperture. This move is desirable to prevent unwanted target charges being induced when the various potentials are applied.

Using the test card an approximately correct picture is obtained by setting target potential at about 2 volts above the cut-off value (see §(c)) and suitably adjusting beam and beam focus. The lens aperture is then adjusted so that picture whites are approximately at the 'knee' of the tube transfer characteristic (see §(c)).

Then, in sequence, the following steps are followed:

(1) Adjust scanning amplitude.

(2) Set camera to look at a uniformly bright surface and adjust beam alignment controls for maximum output consistent with the best minimum shading.

(3) Direct the camera at a test chart and adjust image section controls to obtain minimum 'S' distortion at the highest photo-cathode focus voltage that can be obtained from the camera.

(4) If necessary rotate tube and or the yoke so that the sides of the target and field mesh frames are horizontal and vertical.

(5) Adjust the scanned raster to its correct size.

(6) Adjust decelerator to minimize corner shading and optimize picture geometry.

(7) Re-adjust the lens aperture to obtain desired picture gamma.

(8) Adjust multi-focus for maximum output.

(9) Minimize black shading by capping lens and proceeding as follows:

　　(a) Adjust field mesh voltage to optimize the picture.　Recheck beam focus.

　　(b) Minimize line shading by slight re-adjustment of multi-focus.

　　(c) Finally, apply line and frame shading if absolutely necessary to minimize the black shading and set ' black level '.

　　　　N.B. The practice of correcting non-uniform studio lighting by adjusting the tube shading controls is not recommended as it leads to the need for continued adjustment as and when a camera is panned, and introduces errors in the ' black level '.

(10) Uncap lens and re-adjust:—

　　(a) Beam current to just discharge peak white.

　　(b) Image and beam focus for optimum resolution.

　　(c) Camera gain control for required video signal output.

It is convenient to discuss the properties of the picture, created by applying the above technique, in the sequence already listed.　However, at this point it must be stated that that part of the total system concerned only with display is assumed to have been checked by means independent of the camera so that it is known not to contribute to picture errors. This means that any differences between the original chart and its reproduction must be due to the camera tube, transmission errors also being discounted.

## §(a) *Geometry*

The optical system projects on to the photo-cathode of the camera tube a precise, naturally coloured image of the original scene.　A geometrically precise electron image leaves the photo-cathode plane with all colour and brightness information translated into electron density according to the spectral response of the photo-cathode.　It is upon traversing the electron optical lens between the photo-cathode and target that the first distortions are introduced.

Due to the different emission energies of the photo-electrons not all of them are capable of being focused simultaneously in the target. Non-homogeneities in the electrostatic and magnetic fields cause trajectories to vary in length according to the position in the image. The resulting image on the target is consequently not perfectly rectilinear. In early tubes it was arranged that the distortions were compensated in the scanning section but any move towards perfection could be approached only by improving the performance of both the image and scanning sections of the tube.

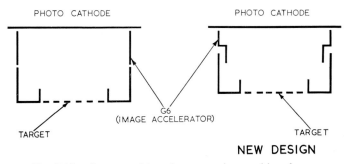

Fig. 7.11.   Image orthicon image sections—old and new.

In order to produce a rectangular photo-electron image at the target, theoretical and practical investigations have ensured that current tube design is satisfactory.   One particular deficiency of early tubes was their inability to reproduce a single image of a single bright object.   A second or ghost image was produced from photo-electrons being reflected from the target and returning to it at high velocity after a second reflection at the photo-cathode.   The second point of incidence at the target was different from the first and so a second image was formed.   Figure 7.11 shows sectionally the original design and that which eliminates the

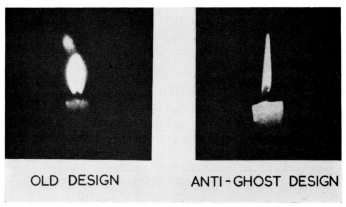

Fig. 7.12.   Candle—ghost and no ghost.

primary ghost image. Figure 7.12 shows television pictures of a candle flame reproduced by each type of tube.

In the scanning section, the second source of geometrical distortion, the main improvement came with scanning coil design. Even so, the fundamental impossibility of terminating a magnetic field abruptly, forces an empirical compromise. Beurle[42] has analysed the problem to show that distortions are minimized if the reading beam electrons enter and leave the scanning field with similar directional attitudes. No scanning can produce an accurate deflection field except in a homogeneous medium. It is therefore necessary to ensure that magnetic losses in the metal beam focusing electrode applied to the tube wall are peripherally constant. It is sufficient to evaporate this coating to a uniform thickness. The residual magnetic aberrations of the deflection field are proportional to the deflection angle and hence are greatest around the edges of the scanned raster. These effects are intensified by the inevitably non-uniform electrostatic end field of the beam focus electrode.

Two antidotes are possible. First an annular electrode—termed the decelerator—is mounted between the beam focus electrode and the target and is connected to a separate potential supply. Alternatively or additionally a highly transparent mesh is mounted parallel to the target and roughly in the end plane of the beam focusing electrode[43]. The exact location of this so called 'field mesh' is determined by the scanning beam diameter. The latter varies as the beam electrons travel between the gun anode and the target due to the inevitable spread of initial emission velocities. Controlled by the values of the electrostatic field due to G.4 and that of the main focusing solenoid the beam cross section consequently varies in a regular manner. Several nodal points usually occur and it is, of course, one of these which is focused in the target plane to ensure good resolution. The field mesh is located at a beam antinode in order to minimize interference effects. Unfortunately the beam's incidence upon the field mesh gives rise to secondary electrons and these will mix with the beam electrons and, eventually, the output signal. Since these secondaries are of a slightly lower velocity from the wanted signal their arrival at the output terminal will be delayed and their signal will be displaced on the received picture slightly displaced from its correct position. This disturbing second image was eliminated by Hendry[44] who suggested the inclusion of a cathode biased short cylinder in front of the gun anode to prevent the passage of the unwanted secondaries to D.1. This suppressor electrode was later made of asymmetric length in order to deflect the return scanning beam away from the exit aperture in the gun anode so as to prevent its appearing in the reproduced picture.

Geometry errors are usually measured as the displacement of any picture point from its true position and are expressed as a fraction or percentage of the picture height or width.

## §(b) Resolution

One feature of resolution has already been mentioned in §(a) in discussing the scanning section of the image orthicon. Probably the greatest contributory factor to good resolution is to have an image section working at optimum efficiency.

Beurle *et al.*[45] have analysed the unavoidable deteriorations associated with establishing the charge image to be read by the scanning beam. In addition to these, practical design problems exist. It has been previously stated that the spectral energy distribution of photo-electrons is not constant. In the image orthicon, and other image tubes working with the whole visible spectrum, this means that the photo-electrons will not have a single focusing potential. For example the ' red ' and ' blue ' electrons need values different by about 3 volts. A compromise is therefore necessary but it is clear that the difference is less significant as the mean focusing potential increases. Hence, as high a value as possible is used. This has the added advantage of increasing the secondary electron yield at the target and so enhancing sensitivity. A further benefit from using high velocity photo-electrons is the minimizing of dispersion by mutual repulsion.

The sharpness of the electron image at the target can be seriously affected by scanning field leakage, and it is necessary by strategic screening and positioning of the scanning coil assembly to minimize this trouble. Some camera manufacturers have included an auxiliary scanning yoke[46] around the image section of tube. This yoke is fed by the scanning currents but in reverse phase so that any unwanted field is destroyed by an equal and opposite one.

The effect of the target mesh upon resolution has not yet been completely determined. To a first order, however, it is clear that the mesh should cause as little obstruction as possible to the photo-electron stream and also be of such a pitch as not to produce interference patterns with the scanning line periodicity. Mechanical limits impose a disturbingly low maximum transparency for this component.

The target membrane itself is another most important factor controlling resolution. Its purpose is to receive the photo-electron image and store it between scans without degradation. Of necessity the target must possess volume conductivity and transverse merging of charge is inevitable. It is thus very important to regulate the specfic resistivity and thickness of the target material to fine limits about an optimum value[47].

Also inevitable is the exposure of the two surfaces of the target to the processing atmosphere of photo-activation—basically conducting—and of one of its surfaces to the control mesh.

The first of these is controlled during tube processing by keeping the target as hot as possible to prevent condensation of the various metal vapours and also by treating its surfaces with chemically resistant materials such as highly stable oxides. Possible transfer of mesh material—usually soft copper—either by sublimation or contact transfer (the target and mesh often sag and touch during the main vacuum bake) is minimized by coating it with a metal possessing a high melting point.

Resolution is also obviously much affected by the sharpness of the scanning beam and it must be the aim of the tube designer to ensure maximum efficiency in this area. The beam should be composed of electrons each having the same energy, i.e. ' monochromatic ', so demanding an efficient emissive coating to the thermionic cathode to ensure the minimum velocity spread among the beam electrons. This also helps to improve signal to noise ratio as will be seen later.

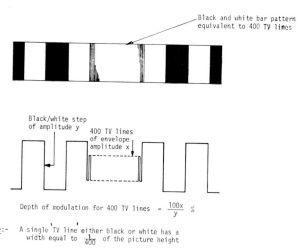

Fig. 7.13. Diagram illustrating method of measuring resolution.

So far in this section we have considered only the pure physics of resolution. There is another side to the subject namely ' sharpness ' or subjective resolution.

We have postulated that the target must store discrete charges. If this is so then between the charges or, perhaps more precisely, between picture elements there must exist capacitance. This means that information edges will have a disproportionately large charge storage capacity. The picture information will thus tend to have emphasized outlines and appear ' sharper '. The outline will be relatively finer as the main information becomes larger with the result that the larger target of the $4\frac{1}{2}$ in. tube gives ' sharper ' pictures than are obtainable under similar conditions with the 3 in. tube[48].

Resolution is usually expressed as the depth of modulation of a particular line frequency, usually 400 TV lines†, compared with a single black/white step (fig. 7.13).

† Equivalent to 533 picture points per picture width or ' points '.

## §(c) Gamma

Picture gamma is a measure of the half-tone or grey scale reproduction in the finished picture. It is often expressed as relationship between the voltage inputs to the reproducing cathode ray tube corresponding to the optical input to the camera from a specified range of grey tones from black to white and is consequently measured on a voltage oscilloscope (fig. 7.14).

The topic of picture gamma is an extremely controversial one due to its subjective nature, dependence on viewing conditions and personal preferences of the viewer, hence the decision in this book to adopt ' an unspecified, opinion free test chart '. The treatment is also restricted to include only those factors in the image orthicon which affect the reproduction of grey tones—from black to white. Effects due to the selective colour sensitivity of the photo-cathode can be ignored since the test chart contains no colour. For convenience the contribution of the reproducing cathode ray tube and its ambient viewing conditions will also be omitted. Each is extremely important when considering overall system gamma from a personal viewpoint.

The photo-electric conversion at the photo-cathode is linear. At the target each incident photo-electron produces an instantaneous proportionate positive charge. Unfortunately some of this charge is destroyed before it can be evaluated by the scanning beam, and although the precise electron dynamics at the target are not known, the following account is supported by most of the observable facts.

The image orthicon target is essentially an insulator and electrical contact to it can be made only by the photo-electrons or the scanning beam electrons. With no light input to the tube the latter will cause the target to stabilize at a potential below that of the thermionic cathode by an amount equal to that of the most energetic beam electrons—very nearly cathode potential. With light input, secondary electrons will be produced at the target and will travel to the target mesh if the latter provides a positive attractive field. The mesh potential is made adjustable and if a sufficiently negative value is chosen, no target secondary electrons can be collected and no positive charge can be stored and consequently the scanning beam is again prevented from reaching the target. As the mesh potential is raised, a value is reached at which the beam just begins to land and a video signal is produced. This value of the mesh potential is called ' target cut-off potential '. At mesh potentials above this value normal video signals can be obtained.

Consider an arbitrary potential $V$ above the cut-off value. As the secondaries leave the target its potential rises and continues to do so until it reaches a value sufficiently positive, $V_s$, that the mesh now presents a negative field gradient to the secondaries and further collection is prevented. A stable positive target potential is established. Diagrammatically this may be represented as in fig. 7.15. It may be

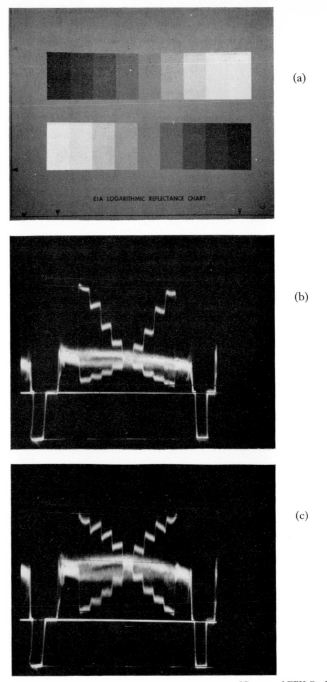

[*Courtesy of EEV Co. Ltd.*]

Fig. 7.14. Photograph of grey scale and oscilloscope traces showing (b) exposure to the knee and (c) exposure for operation. (See para. (e) page 96.)

considered that the capacitance, $C$, formed by the target and its mesh, is now fully charged and that the familiar relationship exists, viz.

$$V_s = \frac{Q}{C}$$

where $Q$ is the charge created by the input photo-electrons.

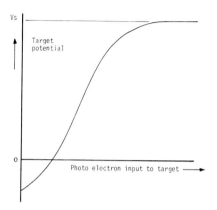

Fig. 7.15.   Relation between potential of an image orthicon target and its photo-electron input.

From this expression it is seen that (1) $V_s$ increases as $C$ decreases for constant light input.   This means sensitivity increases inversely with target capacitance—see later.   (2) For constant $V_s$, $Q$ increases linearly with $C$.   Interpreted this means that a greater range of $Q$—i.e. from zero to a larger maximum value—can be accommodated when $C$ is high.

In the context of the present discussion a greater contrast range of light input can be linearly accommodated with a high capacitance target. It is important to appreciate the significance of the word 'linearly'. As fig. 7.15 shows when $V_s$ has been established no further rise in potential occurs but with increased photo-electron input, signal output is maintained and so it may be argued that such a condition represents infinite contrast range capability.   This is true and for an added reason. The value $C$ has been ascribed to the straight capacitance between mesh and target.   At information edges however (fig. 7.16) as seen earlier, an additional capacitance exists between the charged target element and its uncharged surroundings—transverse or inter-element capacitance is established.   This means that on the 'whiter' side of information edges a disproportionately high charge can exist and a white square is reproduced with a bright border[49].   If the contrast range of the input light is so high then, although no increased signal is produced from the high-lights,

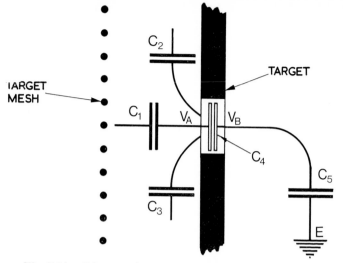

Fig. 7.16. Diagram showing picture element capacities.

intelligence is preserved because all information is outlined by the above described ' edge-effect '.

Edge-effects of course are incorrect and for true reproduction the relative transverse capacity must be minimized. This is achieved by making the main target capacity as high as convenient and maintaining a high ratio between picture element size and inter-element capacitance—i.e. by using as large a target as possible.

A further departure from the simple case described also affects the gamma of the image orthicon. It has been stated that photo-electrons incident upon the target produce target charge and hence video signal. Unfortunately all photo electrons are not incident upon the target. Some are intercepted by the mesh and themselves produce secondary electrons. The latter are attracted by any positive field in their vicinity such as that due to highly charged target areas. Since they are of low velocity the mesh secondaries do not give rise to further emission at the target but merely depress its charge value and signal proportionality is upset. This effect begins to be of significance when mesh and target are at the same potential and intensifies as the target potential rises. The overall result is a tendency to soften the gamma in the brighter parts of the picture dependent on their area. This inconstancy of gamma is probably the greatest objection to the image orthicon type of television camera tube especially for use in colour cameras.

A further modification of gamma occurs as the positive picture charge is transferred through the target. The transfer process is inductive, i.e. dependent upon the relative capacities between the faces of the target and of the scanned face relative to its surroundings. The ratio must be kept as high as possible to avoid dilution of the picture charge. The spacing of field mesh to target is important in this respect. It is also desirable to use as thin a target as possible to maintain its internal capacity as high as possible. The newly developed ELCON† glass which relies on electronic rather than ionic conduction for charge neutralization after scanning is particularly advantageous in this respect[(47)]. Targets made from it have a higher self-capacity than is possible with the original soda-lime material. This derives from a higher dielectric constant and a smaller thickness.

### §(d) Ratio of signal to noise

The term ' noise ' is borrowed from audio technology and applies to the unwanted and annoying background which contaminates or lowers the signal information.

As in electro-acoustics, where any signal producing electric current has a superimposed random variation, so the video output of a camera tube carries a noise component which appears on the reproducing cathode ray tube as a coarse grain structure sometimes with an emphasis or bunching in the line direction. Sometimes called ' snow ' the phenomenon is not to be confused with either the vertical streaking generated when scratched film is being transmitted or the horizontal streaks caused by faulty video tape processing.

Fundamentally, the lower limit of noise in an image orthicon is that associated with the photo-emission. All subsequent processes in the tube must add to this value. However, the magnitude of the scanning beam makes the contribution of the photo-cathode a negligible quantity.

Since the video signal originates as the difference between the scanning beam incident upon and leaving the target this is the main source of noise in the tube. It is essential that the beam be mono-chromatic or as nearly so as thermionic cathode technology will allow. The splitting of the beam at the field mesh and target, although reducing the electron current constituting the output from the tube, cannot reduce the original noise component. At the target the stored charge has an associated noise constituent. In general an electron current, $i$, and its associated noise $\bar{i}$ are related by the following expression:

$$\bar{i} = \sqrt{(2eif)},$$

where $\bar{i} =$ mean value of noise current, $e =$ electronic charge, and $f =$ frequency band-width of the system.

The signal to noise ratio (SNR) then becomes:

$$\frac{i}{\sqrt{i}} = \frac{i}{\sqrt{(2eif)}},$$

$$SNR = K\sqrt{i}.$$

† Trade name of English Electric Valve Company.

93

It is clear that to minimize noise the signal producing current must be as large as possible and, conversely, that low currents will contain a maximum amount of noise. Visually, on the viewing screen, this means that in the darker picture areas one has the greatest annoyance.

Signal current, the difference between the forward full beam and that reflected at the target, is directly related to the charge present on the target produced by exposing the tube to light. Target charge depends on target capacity, target mesh voltage and light input. Of these only target capacity is a design feature of the tube. For high signal to noise ratios tube manufacturers aim to make the target capacity as high as possible. An upper limit is imposed however by other considerations.

Fig. 7.17. Photograph of chequer board to illustrate beam pulling.

To completely utilize a high target capacity one needs a high target voltage and enough light to fully modulate it. Unfortunately, high potentials on the scanning side of the target are undesirable. If there exists scene information such as adjacent blacks and whites, then the high potentials associated with the whites tend to pull the beam towards them from the black area position (fig. 7.17). The result is to diminish the size of black areas and increase the size of whites[49]. The phenomenon is variously known as 'beam pulling' or 'ballooning'. Targets with field mesh deceleration for the scanning beam suffer least from the defect.

A further feature of high capacitance targets is that the charge can sometimes be so high that the beam cannot discharge it in one pass. Several passes are sometimes necessary to remove a highlight completely. This means that a ' persistence of vision ' exists and if the object in the scene is moving a trail is seen.

A similar effect occurs at target charges so low that the scanning beam is unable to approach close enough to discharge efficiently. Again a trail is produced by moving objects.

The scanning beam discharges the target most efficiently when the latter is at some white potential arbitrarily determined for each individual tube by its components.

From previous work it is seen that for a given available scene illumination, maximum beam discharge potential at the target is obtained for low values of target capacitance—subject of course to the mathematical relationship. For this reason, low capacitance targets are said to have high sensitivity. The fact that this is accompanied by higher noise is unfortunate and there is a growing tendency to relate sensitivity to signal to noise ratio.

The annoyance value of noise is its main feature. This cannot be measured but to remove or reduce the subjective nature of its assessment many attempts have been and are still being made to evolve an instrument method for measuring it[50].

Noise has amplitude when displayed on a voltage oscilloscope (fig. 7.18), and it has been industrial practice for many years to ' measure ' this, operate on the value by an empirical conversion factor and use the result as a measure of the noise[51]. The conversion factor has some mathematical basis in converting a double amplitude peak value to an

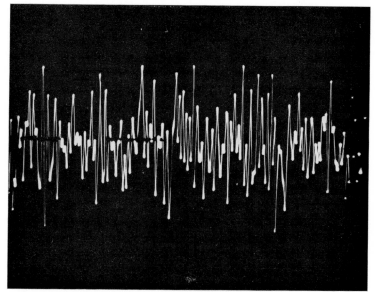

[*Courtesy of Mr. Ray Knight, A.B.C. Television*]

Fig. 7.18. Diagram of typical noise trace.

95

r.m.s. one but, in practice, 'personal' factors were ascribed to the engineer making the measurement. The obvious difficulties ensuing from a multitude of values from a multitude of operators encouraged the search for an absolute method. Attempts to[52] measure the peak to peak values of the noise voltage trace scientifically, reduced but did not eliminate the personal element. The first instrumental suggestion was made by Weaver of the BBC[53]. In this method the blanking signals, unavoidably associated with a television signal and which made a simple direct meter measurement impossible, were removed by a selective tuning process. The tube noise was then directly compared by switching with calibrated noise, thermally generated. Subsequent work by Edwardson[54], and Holder[55] of the BBC and Grosskopf[56] of Institut fur Rundfunktechnik of Munich, has improved the technique and commercial measuring equipment is now available.

§(e) *Sensitivity*

In general, the sensitivity of an image orthicon is directly related to the amount of light incident upon its photo-cathode necessary to fully modulate its target.

The relationship between incident illumination in lumens and the signal current is generally as shown in fig. 7.19. The point of inflexion, called the knee, at which the slope of the curve—or gamma—is 0·5 is

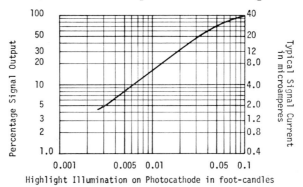

[*Courtesy of EEV Co. Ltd.*]

Fig. 7.19. Relationship between the light input and signal output for a typical image orthicon.

taken as the reference point. Illuminations above the value at this point may or may not cause a significant increase in signal current. Simple theory calls for minimal increase but this is incorrect as has been seen in §(c) earlier.

The photo-cathode illumination at the knee or its reciprocal is taken as the sensitivity of the tube. To determine sensitivity, the tube in the camera is used to view a logarithmic grey step wedge, and the camera lens opened until the voltage waveform trace of the wedge begins to depart from linearity at its highlight end (see fig. 7.14 b). The end white step is then at the knee of the light transfer characteristic and its brightness is used to calculate the photo-cathode illumination.

The following formula obtains:

$$I_s = \frac{I_p 4f^2(1+m)^2}{TR},$$

where $I_s$ = scene illumination in foot-candles, $I_p$ = face plate illumination in foot-candles, $f$ = lens aperture, $m$ = linear magnification from scene to photo-layer, $T$ = transmission of lens, and $R$ = scene reflectance.

As an alternative to 'sensitivity' one might use the term 'detectability' which is unrelated to 'sensitivity' but is indicative of the tube's usefulness in low light conditions.

Clearly 'detectability' is related to noise. In fig. 7.19, if the equivalent noise current is indicated on the $Y$ axis, illuminations above the corresponding light level are necessary if any 'signal' is to be measured or seen.

An additional requirement could be stipulated. At the threshold illumination level mentioned in the last paragraph it would be possible to follow, on the viewing screen, any switching sequence of the light source, At that illumination level it may not be possible to determine if one or more light sources were being used. To include such a provision would make 'detectability' dependent also on resolving power. International technical discussions were commenced in 1967 with a view to settling the question.

There is also some merit in the idea of ascribing an ASA or DIN rating to television pick-up tubes and Neuhauser has explored this possibility[57]. In this regard difficulties will arise, especially in colour where spectral sensitivities and gamma of both film and tubes can vary considerably according to processing.

## §(f) Picture uniformity

Under this heading the general constancy of signal level and appearance of the picture over its whole area are considered.

For this purpose, the picture area is conveniently divided into strips both vertically and horizontally. The strip width is arbitrarily determinable. A display of the voltage waveform along any strip gives a measure of signal level along it and a value for 'shading' can be quoted.

To examine the vertical strips individually a special electronic sampling technique is necessary[58]. In general it is sufficient to use a composite display of all the vertical or horizontal strips superimposed. In this case the width of the trace of the horizontal strips is a measure of the vertical non-uniformity and vice versa.

Small area disturbances are not detectable by the above technique and resort is made to viewing an evenly illuminated back cloth or screen, and panning the camera across it. All minor non-uniformities on the tube's reproducing surfaces can be easily seen.

## §(g) *Ease of setting-up*

The user of a television camera requires that it be a simple machine. Television has passed out of the novelty stage during which getting any picture at all was an achievement! It is now a recognized entertainment medium developing its own production art, and picture making must be a simple process.

One may plot a graph as in fig. 7.20. Curve A is preferred by many operators but Curve B, allowing a smaller degree of flexibility with a higher level of performance, is more acceptable. Tube and camera designers aim to produce such a product and currently available image orthicons come very close to this ideal.

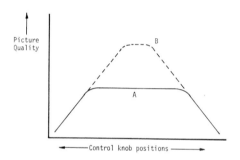

Fig. 7.20. Hypothetical indication of the ease with which a television picture can be obtained.

## 2. *The image isocon*

This pick-up tube, announced in 1949[59], is externally similar to an image orthicon but generates its signal electrons by, and proportional to, the target charge produced by the photo-electrons.

As the reading beam scans the target, in addition to being modulated it causes the production of electron showers at the points of picture charge. These scattered electrons have no preferred direction of emission and the precise mechanism of emission is not yet understood.

In the isocon the video signal differs from that of the image orthicon in that it is directly instead of inversely proportional to the scene brightness. This means that in black picture areas there is no signal and hence no noise and the tube can, in consequence, 'see' into darker regions. Video signals much less contaminated by noise are produced.

In an image isocon it is necessary to separate the scattered 'isocon' electrons from the specularly reflected 'orthicon' beam. The process is simply described but difficult to achieve in practice[60].

98

Since the isocon electrons have no preferred direction of travel and the scanning electron beam has, it is possible to mount electrodes in strategic positions to separate the two streams and to collect only the scattered one. The additional electrode system is shown diagrammatically in fig. 7.21.

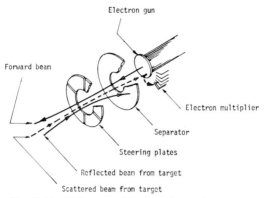

Fig. 7.21.   Diagram of isocon electrode system.

The forward passage of the scanning beam through the plane of the steering electrodes causes it to begin spiralling and it meets the target with an angle of incidence greater than that in the image orthicon. The depleted beam leaves the target at approximately the same angle but on the other side of the normal, and spirals back to the steering plane. Here it undergoes further deflection and is collected on the separator.

The isocon electrons generated at the target have no preferred direction of movement on emission and are merely constrained by the

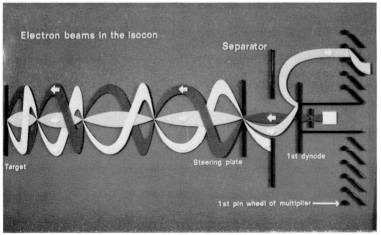

[*Courtesy of EEV Co. Ltd.*]

Fig. 7.22.   Diagram of isocon reading beams—undeflected.

axial magnetic field and follow the electrostatic field back towards the gun. They suffer a slight deviation from the steering field but pass through the separator aperture and impinge on the first dynode of the electron multiplier to become the signal beam. This beam is of smaller magnitude than that in the image orthicon and hence it is necessary to include extra pinwheel dynodes to ensure adequate video output level. Figure 7.22 is a diagrammatic representation of the undeflected beams in the isocon.

The lack of noise in the isocon blacks allows much wider target spacings to be used to increase sensitivity. Consequently, the isocon has inherently a higher overall sensitivity. Table 7.1 shows comparable data, given by Van Asselt[60], for tubes having target spacings of 0·001 inches.

| Parameter | Image isocon | | Image orthicon | |
|---|---|---|---|---|
| Signal to noise ratio† | 30 : 1 | 12 : 1 | 30 : 1 | 12 : 1 |
| Photo-cathode illumination in foot-candles. | 0·002 | 0·0001 | 0·005 | 0·002 |

† Ratio of signal to r.m.s. noise.

Table 7.1. Comparison of image orthicon and image isocon.

Relative to the image orthicon the isocon is at present somewhat more difficult to set-up and to operate. This fact has held back its commercial development but the modern need to extend the range of television into low light applications has encouraged new work[61] to compare the isocon with other low light tubes. The mechanism of beam control is now more fully understood and tubes with new electron optics are in development.

# CHAPTER 2.8

## photo-conductive pick-up tubes

THESE tubes rely upon a photo-conductive material as their light sensitive element. They invariably use orthogonal reading beams, focused and deflected either electrostatically or magnetically or by a combination of both.

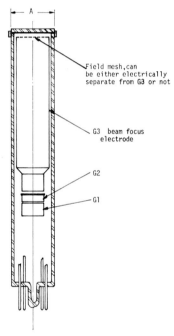

Fig. 8.1.   Diagram of typical vidicon.

Figure 8.1 shows diagrammatically the general form of the tube type. The field mesh can either be separately connected and the diameter, $A$, can vary from 2 in. down to $\frac{1}{2}$ in. for commercial tubes. Experimental samples of 6 mm diameter have been made.

The photo-conductive layer can be either amorphous as in the vidicon series[62], a true semiconductor as in those types similar to the Philips plumbicon[63], or a single crystal as, for example, silicon[64].

The most commonly used material in the first category is Antimony trisulphide, ' $Sb_2S_3$ ', but containing slightly more antimony than the stoichiometric quantity. The true semiconductor tubes use lead oxide possibly doped as discussed later. Up to the time of writing only silicon has appeared as a mechanically divided[65] photo-conductive mosaic but doubtless other materials will emerge as research continues. It is convenient to consider these three categories separately. Silicon will be discussed in Chapter 2.10.

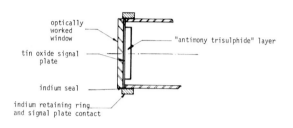

optically worked window

"antimony trisulphide" layer

tin oxide signal plate

indium seal

indium retaining ring and signal plate contact

General arrangement of the photo conductive target of a vidicon. The sensitive layer can have many forms. It can be homogeneously solid or spongy. Alternatively it may be of sandwich construction combining layers of both types laid down in any sequence.

Fig. 8.2.   Diagram of a typical photo-conductive vidicon target.

## 1. *Amorphous photo-conductive tubes*

In the vidicon the amorphous layer of antimony trisulphide is thermally evaporated on to a highly transparent and conducting film of stannous oxide in turn supported by a glass plate usually the end window of the tube. Figure 8.2 shows this arrangement. Light incident upon the photo-layer through the window and tin oxide film causes local reductions in the resistance of the antimony trisulphide. A potential applied to the signal plate can now leak selectively through the photo-conductor to form a potential pattern on the vacuum side of it and partially discharges the capacitance between the signal plate and the scanned surface, which is, in the dark, stabilized at cathode potential.

When scanned, each picture element is returned to cathode potential and the recharging of the elemental capacitance produces a pulse in the signal plate. This, passing through the potential supply lead, produces the video voltage output from the tube across the signal resistor.

With simple hard vacuum photo-conductor deposits the elemental capacitance is high and produces low voltage patterns on the scanned side and the discharge efficiency of the beam is low. Lag, due to incomplete pattern erasure, results. To reduce this undesirable property, the signal plate potential may be increased but this allows the inevitable non-uniformities in the thickness of the layer to produce uneven signal production. The more acceptable remedy is to increase the voltage swing by evaporating a thicker layer of photo-conductor to reduce the elemental capacitance. To avoid the consequential high resistances of thicker hard films the material is evaporated in a gaseous atmosphere[66]. A thick porous film is formed in this way. Its back to front resistance is much greater than that of the hard thin film so reducing the associated dark current. Sandwich combinations of hard and porous films can also be utilized[67]. According to the order of deposition on the signal plate base so the sensitivity, spectral response, and speed of response changes.

A a general rule thin, hard layers of antimony sulphide have high and predominantly blue sensitivity and high lag. The porous layers, on the other hand, have lower white sensitivity but relatively higher in the red spectral regions; they also have lower lag.

Lag in an antimony trisulphide layer arises from capacitance effects between the array of picture elements and the backing signal plate and also from delays in the creation and recombination of the conducting centres. Lubszynski[68] has analysed the separation of the two effects. Briefly, it can be said that the main contribution to lag comes from the photo-conductive effect. Discharge lag is largely controllable as a function of the layer thickness and the potential at which it is operated.

Simple explanations do not yet exist for the various phenomena described but the reader is referred to original work for details of the researches so far undertaken.

## 2. *Semiconducting photo-conductive tubes*

One basic material for the photo-sensitive layer of these tubes is yellow lead oxide. Its use was described in a report by Heinje in 1957[69]. Since that time its properties have been investigated in detail and its capability of taking up minute amounts of Tl or Cu to form a true n-type semiconductor, have been discovered.

The following is a simplified account of the mechanism by which the complicated layer of special lead oxide is believed to function. Again, the student is recommended to refer to the original papers[63] on the subject for further detail.

The photo-sensitive layer of lead oxide behaves as a mosaic of p–i–n diodes. Backed by a fully conducting film of tin oxide and having as the scanned surface a complicated layer of specially treated lead oxide, each diode is effectively bounded by blocking contacts. The tin oxide is n type while the special film is p type.

When illuminated, electrons and holes are created in the intrinsic central layer of the sandwich. These electrons and holes are separated by the applied field and a charge pattern is established on the inner surface to be scanned in the conventional way for signal production.

Because of its strict semiconducting nature the lead oxide layer has virtually no dark current, so enabling high signal plate potentials to be used allowing high operating sensitivity without introducing picture shading. The spectral sensitivity is good although perhaps a little deficient above 6500 Å in tubes with minimal lag. Extension of the threshold wavelength above this value does tend to increase the lag. The gamma of the material is unity and this is regarded as a disadvantage by the many observers who prefer a limiting type of light transfer characteristic.

Camera tubes incorporating the lead oxide layer are finding successful applications for colour television cameras.

### 3. *Signal generation*

Usually, as has been indicated in the preceding text, the video output is taken from photo-conductive camera tubes as current changes in the signal plate which backs the photo-conductive layer. A simple circuit is shown in fig. 8.3.

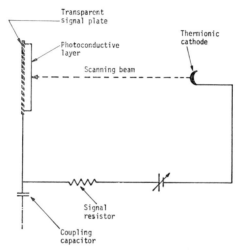

Fig. 8.3. Typical video signal circuit for a vidicon pick-up tube.

Alternatively, the return beam read-out system may be employed. In this latter case an electrode arrangement similar to that used in the image orthicon is utilized.

### 4. *Noise*

Since the signal in a conventional vidicon type tube contains none of the components of the generating scanning beam its noise to signal ratio is very low. This is probably considered to offset its other disadvantages relative to tubes with return beam signal generation.

The minimal noise associated with direct beam read out facilitates the electronic adjustment of signal characteristics such as gamma, resolution and unsatisfactory spectral response. This has led to the widespread adoption of photo-conductive tubes for colour television cameras where signals having precise mathematical accuracy are required for correct colour reproduction.

# CHAPTER 2.9
## miscellaneous camera tubes

In this chapter those camera tubes which have yet to see widespread adoption will be considered. The general feature of the group is that of higher sensitivity. Such sensitivity is unnecessarily high for conventional broadcasting where aesthetic tastes demand lighting of a certain intensity and mood to satisfy the programme requirements.

The television process of remote viewing is of use in many industrial, military and medical areas where the only requirement is accurate observation. For these and other needs tubes with extra detecting power are required while still retaining the frame storage principle.

Using the signal generating techniques described in the last chapter it is necessary to introduce an amplifying component into the operating mechanism of the tube. The choice rests between intensifying the charge input to the target or increasing the sensitivity of the target itself.

Means of increasing the photo-electric conversion process of transparent photo-cathodes have been described[70]. In general they rely upon increasing the fraction of input light utilized. This can be achieved by roughening the inside surface of the entrance window—the base for the photo-cathode—either in a regular pattern similar to a microscopic Fresnel lens, or randomly as for a conventional diffusing surface. The disadvantage of such a practice is the possible effect on resolution. Other suggestions[71] involve the incorporation of interference reinforcing substrates into the photo-emitting layer itself.

Perhaps an easier approach is to arrange an intensification of the electron image either in or separately from the tube containing the storage target.

Tubes containing an image section such as the image orthicon are fairly easily modified to incorporate one or more stages of image intensification. The method described by Morton[72], among others, is to mount a transmisson type secondary emission membrane between the photo-cathode and target (fig. 9.1). This may be any of the types mentioned in Chapter 2.12.

Alternatively a phosphor/photo-cathode sandwich may be used as an intermediate multiplying stage[73]—see Chapter 2.12. A separate tube containing only a photo-cathode and phosphor may also be used in front of the photo-cathode of the scanning tube.

Perhaps less difficult once it has been made is to use a target of higher sensitivity in those tubes already using an image section. Sensitivity in this sense is used as a measure of charge stored relative to photo-electron input.

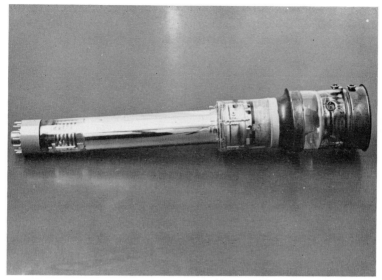

Fig. 9.1.   Photograph of image intensifier orthicon.

Two types of such target have been described, firstly in which conductivity is induced by the photo-electrons and secondly in which, by virtue of the internal secondary emission effect, a much larger stored charge is produced.

A typical example of the first category is the EBICON—similar in concept to the image orthicon but, in one form, using a dielectric target of $As_2S_3$[74] (fig. 9.2).   The target is supported on a layer of aluminium oxide, which carries also a thin layer of aluminium to serve as the signal plate.

The photo-electrons are electrostatically imaged on to the target and are accelerated through a potential difference of up to 17 kV; the acquired energy causes them to penetrate the films of aluminium oxide and aluminium to induce conductivity proportional to their intensity in the dielectric.   As in the vidicon, the scanned surface thus takes up a corresponding charge pattern which, when discharged by scanning, gives rise to current pulses in the aluminium film and hence, a video signal.   Dependent on the type of target and its operating condition, very high gains have been realized[75].

While the standard ebicon uses direct read out it is possible to use the return beam or the scattered electrons for signal generation, the

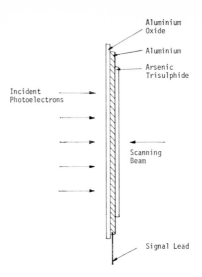

Fig. 9.2. Schematic diagram of Ebicon type target.

noise content of the signal being dependent upon which method is used to generate it.

The second type of amplifying target is similar in its basic concept inasmuch that it is supported on a film of aluminium oxide and carries an

[*Courtesy of Westinghouse Electric Corporation*]

Fig. 9.3. Photograph of SEC vidicon.

108

aluminium signal plate. The target material, however, is potassium chloride deposited in a porous form. The open structure of the KCl layer allows a type of Townsend avalanche to be produced by an incoming photo-electron of adequate energy. The resulting high secondary electron currents can give rise to target gains of up to 500 times. The process is known as secondary electron conduction (SEC)[76]. The video signal can be taken directly from the aluminium signal plate or by return beam modulation as above.

Tubes incorporating SEC targets have one serious drawback and have yet to see widespread adoption. This is due to the relative ease with which the target may be burned if the tube is exposed to a very intense small area light. However, for surveillance purposes such a tube, called the SEC vidicon[77], has been extensively used. Figure 9.3 shows a photograph of one version of the tube.

# CHAPTER 2.10

## solid state systems

IN this chapter we shall deal with picture transducers at present in a very early stage of development, which comprise some form of a planar array of photo-diodes.

While detailed information is not yet available or is beyond the scope of understanding of this book, it is useful to describe one or two directions in which research is developing. Especially worthy of note is the tendency, parallel to that of the transistor in receiver philosophy, for the vacuum tube to be eliminated in picture pick-up systems.

The most recent device to be announced is a vacuum tube identical in appearance to a conventional vidicon and capable of existing in all its known mechanical forms. Developed firstly by Bell Laboratories[64] in the U.S.A., the tube differs from the vidicon in its target. No *in situ* vacuum processing is required. The target is an array of photo-diodes prepared by the new production techniques of micro-integrated circuits. The photo-diodes are conveniently made of silicon.

The Bell tube is suggested for attachment to normal telephones hand-sets so that subscribers may communicate visually as well as aurally. Another tube, intended for detection of infra-red sources such as gallium arsenide light emitters, has been announced by Texas Instruments and named the Tivicon[78]. The target of each tube is basically a single crystal slice of n-type silicon but transformed into a mosaic of photo-diodes by complicated vacuum and chemical processing[65].

The first step in the transformation crystal slice is to form on one surface a layer of silicon dioxide. Then by photo-chemical etching this is selectively removed to form an array of holes. By a diffusion process the silicon is caused to become p-type at the bottom of each hole. An annulus of gold is then evaporated to surround the mosaic so formed and act as the base contact. Finally the silicon wafer is reduced to minimal thickness by further chemical etching and the completed target mounted mechanically into its vacuum envelope. In operation the tube behaves much as does a normal vidicon (see fig. 10.1).

When scanned with no light incident on the target the p-type islands will take up electrons until each assumes cathode potential. As soon as light is incident upon the layer, current carriers (holes) will be created and the p portion of the diode will rise in potential by an amount proportional to the light intensity. Upon recharging by the beam a pulse is generated in the video resistor in the usual way.

The quantum efficiency of silicon arrays can be as high as 60% to give very high sensitivity but other picture producing attributes are completely dependent upon the mechanical and chemical processes involved in manufacture. Resolution is clearly a function of the transverse insulation of the mosaic and blemishes can more easily result from the necessary handling movements.

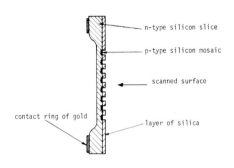

Fig. 10.1. Diagrammatic cross-section of a silicon target.

In the non-vacuum range of devices the scanistor was among the first to be described[79]. Others of similar basic format but of varied techniques for read out have also been announced[80].

In the scanistor type of device the image receiving plate is, in essence, a sandwich of p-type silicon between two n-type layers—the whole forming a continuous array of photo-diodes. Figure 10.2 serves for descriptive purposes only. To each outer surface of the n-type layers is applied, at mutually perpendicular edges, a pair of contact strips.

In one form of the device, across each pair of contacts is applied d.c. bias voltages while additionally to one pair is fed a saw tooth voltage of a similar peak value. Under these conditions a diagonal voltage line is continuously swept across the panel. The current from one outer surface to the other will be constant unless the panel is illuminated when, due to photo-excitation, the passive state will have changed. By monitoring the current in the voltage sweeping circuit and applying well-known detecting techniques, the illuminated state of any point on the panel may be measured.

The practical realization of devices such as the scanistor involve many extremely intricate processes on which much further work will have to be done before they become of practical, everyday application. In the broad field of television, Weimer and others of RCA have described[82] a device of advanced design which is also intended for self-scanning. The device, as yet unnamed, combines the principles of the silicon vidicon and the scanistor principle.

In the silicon vidicon some elemental capacity is essential for charge integration or storage and so it possesses the attendant undesirable features already described. In the new device excitation storage is employed. This principle is possible only with photo-sensitive materials having high internal gain in which the photo-electric effect is stored as excited carriers. The life time of these carriers is ideally the same as the scanning period. High quantum yields are possible with such systems utilizing materials such as cadmium sulphide, selenide and combinations of the two.

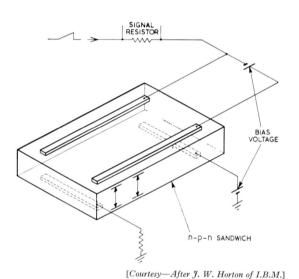

[*Courtesy—After J. W. Horton of I.B.M.*]

Fig. 10.2.   Diagrammatic representation of a scanistor type target.

In the Weimer device the diode switches necessary for self-scanning are incorporated into each photo-conductive mosaic element. This is achieved by the application of dissimilar contacts to them. One electrode is ohmic while the other performs a blocking function. That is, suitably biased it can only allow passage of current when the photo-conductor is in an excited state, i.e. when it is illuminated.

For the ohmic contact, indium or aluminium is used and for the blocking contact, tellurium. This combination of indium, cadmium selenide/sulphide, and tellurium only conducts when the latter is positively biased. It is useful to study the sequence of manufacturing steps in order to understand the method of operation. Firstly the photo-conductor is evaporated on to the substrate through a mesh to form a mosaic of islands (fig. 10.3). Indium contact strips are now deposited to connect these together in rows (fig. 10.4). The indium strips are connected together by a gold bus bar. The anode contact to the diodes is evaporated, again in strips, at right angles to the indium cathode grid. Obviously these must be insulated and the blocking contact also formed.

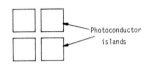

Fig. 10.3. Stage 1 in the preparation of a self-scanned photo-sensor target.

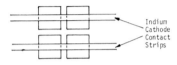

Fig. 10.4. Stage 2 in the preparation of a self-scanned photo-sensor target.

To allow this operation the cathode strips are covered by an insulator such as calcium fluoride in such a width not to obscure the photo-conductor islands (fig. 10.5). The blocking tellurium is now evaporated and, finally, a gold bus contact connecting them together. Each photo-conductive island now has the form of fig. 10.6, effectively divided with a diode in series with each. Figure 10.7 shows a diagrammatic view.

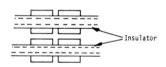

Fig. 10.5. Stage 3 in the preparation of a self-scanned photo-sensor target.

The separation of cathode and anode, governing the resistance of each photo-conductive element, is determined only by the width and placing of the insulator. The required accuracy is readily obtained by mechanical means.

The diode switch formed at each half element is normally 'off' but can be opened by the arranged coincidence of two pulses, one each generated in the anode and cathode circuits.    These can be provided by conventional means or as in the present case by a solid state generator integral with the main sensor array.

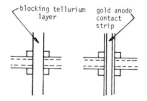

Fig. 10.6.    Stages 4a and 4b in the preparation of a self-scanned photosensor target showing one evaporated element only.

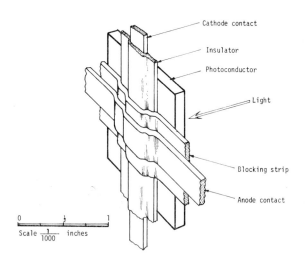

Fig. 10.7.    Perspective representation of one evaporated element of a photo-sensor target.

For the sake of completeness this chapter is concluded with a brief mention of two image conversion processes in which the object information not only cannot be seen with the naked eye but where the detector or camera responds to wavelengths far removed from the visible spectrum.

The first system relies upon the infra-red radiation from the object. The ' object ' can be a human being when the technique of thermography is used to plot a thermal map of the body, or a particular part of it, to an accuracy of 0·08°C.  Such a map will indicate local concentrated areas of temperature differences which are useful in the diagnosis and treatment of such ailments as breast cancer, appendicitis, skin burns, frostbite, etc.

The detector in this case is an extremely sensitive photo-conductive cell giving an electrical output capable of electronic processing and subsequent recording on photographic film[83].

No commercially available pick-up tube is sensitive in the necessary 10 microns wavelength band. For this type of work, therefore, resort is made to photo-conductive detectors such as mercury doped germanium and indium antimonide which do have sensitivity in this region. In conjunction with an efficient optical system such photo-conductive cells can be mechanically scanned over an object to yield a ' picture '.

The second system[84] utilizes a camera tube which has a piezo-electric crystal as its face-plate. An ultra-sonic image of the object is focused upon this window and the scanning beam reads the corresponding voltage changes in the normal way.

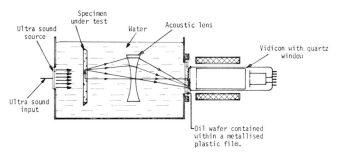

Fig. 10.8. Diagram of the essentials of an Ultra-sound camera.

The system is ideally suited to the non-destructive inspection of possible defects in metals—those that have been welded for example. It is further suggested that some areas of medical diagnosis will benefit from its use. Figure 10.8 shows a diagrammatic representation of a typical system and is self-explanatory as to its method of working.

# CHAPTER 2.11

## light emitters

In this chapter we deal with the reverse of photo-electricity, the phenomenon of 'uminescence or the emission of light as a result of some type of electron stimulation. The light emitting diode is excluded as it is considered to be outside the scope of the present treatment.

The most general transducer in this category is the phosphor. A phosphor is a material which emits light as a result of electron bombardment and is an essential feature of image convertor tubes.

The general emission of light is covered by the term ' luminescence '. The phosphor is said to ' fluoresce ' when under electron bombardment. Any light emission after the electron excitation has been removed is termed ' phosphorescence '[85].

Some phosphors will emit visible light when subjected to ultra-violet radiation, others will emit yellow when irradiated with violet light. These materials are true frequency converters.

The detailed mechanism by which phosphors function is not fully understood and only general explanations of the many experimental phenomena are available.

Just as heat will cause incandescent materials to generate light so will luminescent materials emit light when subject to electron bombardment. The two phenomena are not unrelated.

With an incandescent material, electrons in the atom are raised to an unstable energy level by thermal excitation. In luminescent materials excitation is by electron energy. In each case, as the excited electron returns to its stable ground level, light is emitted. As in atomic spectra the luminescent light usually consists of a band of frequencies. The frequency band is a function of the material under bombardment and it follows that by strategic selection the colour of the emitted light may be determined.

Luminescent materials are crystalline and their atomic structure has the familiar lattice pattern—sometimes containing imperfections due to intentionally included impurities or ' activators '. Within the lattice there are electron traps and emission centres. An electron in an excited state may oscillate between two higher energy levels or between a higher level and the ground state. The return of an electron to its ground state is often delayed by its being held temporarily in an electron trap to cause delayed emission of light or persistence.

As with photo-conductors, the presence of impurity atoms in the lattice can so aid the process of electron transfer to increase the intensity

116

of the emitted light and its colour.    These impurities are called activators and are chosen to improve the efficiency of the phosphor and to control its luminescent colour.

Impurities are also used to control the duration of the luminescence. Nickel for example is commonly used to produce phosphors with the shortest persistence.    With a conventional interpretation, impurities can also be contaminants in the sense of reducing efficiency and causing incorrect colour production.    Iron is a commonly experienced contaminant.

Of the various materials in the phosphor category zinc sulphide is one of the most commonly used basic chemicals.    It is capable of 'activation' with silver to give a blue luminescence and with copper to produce green.    The persistence of the green phosphor is longer than that of the blue.    The formation of the phosphor material is done at high temperatures and according to the temperature used hexagonal or cubic crystals will be produced.    It is hence seen that the possible combinations of activator, temperature of formation, etc., can give rise to a range of luminescent colours from blue through green to orange as the predominant wavelength.    A further possibility is that of mixing phosphors in the screen making process to make other colours, that is by mixing the final colour in the viewing eye.    'White' for example is made by combining yellow and blue, as can be seen by examining a normal television receiver screen with a lens.

Phosphor screen production is usually by a process of allowing the phosphor crystals to settle out of a suspension in water and using some ionizing agent such as sodium sulphate.    Chemical binders are combined with the suspension or added to the screen just after decanting the clear water at the end of the settling period and before drying off.

To prevent burning under intense bombardment and to control the working potential of the screen, an aluminium film is applied to the bombardment side.    As a support for this film and to prevent it from filling the interstices of the settled crystal layer, collodion or some other organic stuff is floated over.    Subsequent baking either before or during processing of the screened tube removes the collodion.

The light output of a phosphor depends on the current input and density, its operating potential, and in intermittent systems, such as television, the duration of the excitation.    Within the limits of this book, however, consideration of scanning systems will be omitted.

The luminance $B$ of a phosphor screen is related to its current input $i$ by the formula:

$$B = K_1 i^\gamma \qquad (11.1)$$

for a constant operating potential.    $\gamma$ is usually not greater than unity (fig. 11.1).

If the operating potential is $V$, then the luminance depends on $V$ according to the equation:

$$B = K_2(V - V_0)^n, \tag{11.2}$$

where $V_0$ and $n$ are peculiar to the particular phosphor.

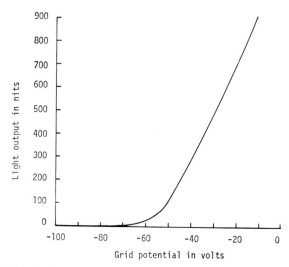

Fig. 11.1. Relationship between control grid potential and screen brightness for a typical cathode-ray tube.

For most applications in this book constant working potentials and continuous excitation are employed. A phosphor screen, within these limitations, is a means for detecting electrons usually originating from a photo-cathode and forms an important component of the conversion tubes described in the next chapter.

# CHAPTER 2.12

## image convertors and intensifiers

In this chapter we shall consider briefly the family of vacuum tubes which contain in one vacuum envelope a combination of photo-cathode and phosphor or other electron detecting media. The simplest form of such a tube is shown in fig. 12.1.

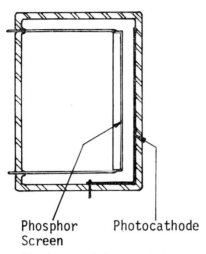

Phosphor Screen       Photocathode

Fig. 12.1. Simple type of electrostatic image convertor.

The photo-cathode and phosphor are in sufficiently close proximity that, using an electrostatic field of approximately 1000 volts per millimetre of spacing, an optical image projected on to the photo-cathode is reproduced as a fluorescent image on the screen. The photo-cathode can be made sensitive to invisible light, such as infra-red, and hence the tube can be used as a light convertor.

Mechanical difficulties are encountered with this type of tube construction inasmuch that no space is available for locating the activating materials for the photo-cathode.

There are two solutions to the problem. Firstly, the separation of the cathode and screen may be increased to a mechanically convenient distance or secondly, the photo-cathode can be formed separately elsewhere and then moved into position.

If the cathode-screen distance is increased direct transfer focusing cannot be relied upon since linear electrostatic fields cannot be maintained. Some form of focusing lens system must be used. This can take the form of a magnetic field from an external solenoid or an electron lens formed by auxiliary internal electrodes.

For convenience the all electrostatic arrangement is usually chosen because this allows a design which additionally produces intensification by concentrating the electron image. A typical tube is shown in fig. 12.2.

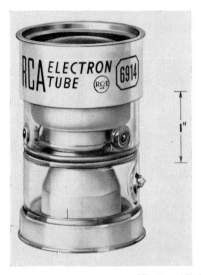

[Courtesy of Radio Corporation of America]

Fig. 12.2. Photograph of an RCA type image intensifier.

Inlet and output windows of these tubes can be chosen to suit particular requirements. For example, quartz for ultra-violet use or fibre optic plates for direct coupling in tandem arrangements.

The simple principle of image convertor type of tube has been developed to a very high order of performance by McGee and his co-workers as the spectracon[86] a name related to its main application so far the observation of stellar spectra (fig. 12.3).

The spectracon is unique in many ways. In particular, its photo-cathode, of narrow rectangular shape, is not formed in the main tube. Its output window is mica of micro-thickness through which the electron image passes for direct recording on electron sensitive celluloid film. The elimination of a final visual image leads to very high resolving powers. Values of 90 line pairs per millimetre are not uncommon.

The exclusion of the photo-activation process from the operating vacuum was considered necessary to avoid contaminating the internal insulating surfaces with the sensitizing alkali materials. A further precaution is the provision of an inner skin of high resistance glass to minimize spurious electron emission otherwise experienced from the glass envelope, a case where secondary emission is unwanted.

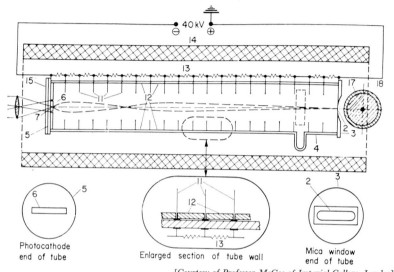

Photocathode
end of tube

Enlarged section of tube wall

Mica window
end of tube

[*Courtesy of Professor McGee of Imperial College, London*]

Fig. 12.3. Section of the Spectracon.

To ensure a uniform accelerating field along the tube regularly spaced metal annulli are provided, each connected to a corresponding supply point of an external potential dividing network. The whole tube and the potential divider is encapsulated in a corona suppressing plastic and is operated immersed in a solenoidal magnetic field or a cylindrical magnet. Focusing of the electron image is achieved by setting up the correct relationship between the magnetic and electric fields.

Considerable ingenuity is exercised in the design of the mica exit window to optimize electron transmission and minimize breakage from such a fragile film and also in the applicator used to ensure intimate contact between the window and the recording emulsion.

The tube operates with a potential difference of some 40 kV in a magnetic field of 160 gauss. Accuracy and stability of these fields are required to ensure adequate image geometry and focus.

Intensification of a high order can also be obtained from tubes in which multiplying stages are included between the photo-cathode and electron detector, usually a phosphor. They can be of the transmission secondary emission type when such materials as potassium chloride are used.

Alternatively, in cascade image intensifiers phosphor/photo cathode sandwiches are used[87]. These consist of thin glass plates on one side of which is deposited a phosphor screen and on the other a photocathode is formed. Figure 12.4 shows such a tube.

(a)

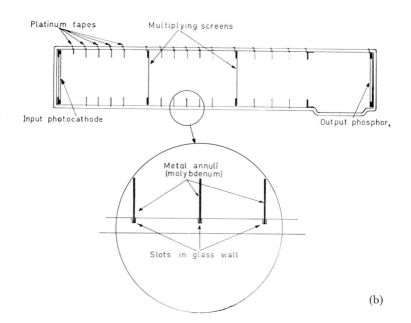

(b)

[Courtesy of Professor McGee of Imperial College, London]

Fig. 12.4. Photograph and section of a cascade type image intensifier (a and b).

The simultaneous processing of several photo-cathodes in the same vacuum is not easy and cascade tubes have found little practical application up to the time of writing.

A tube of the transmission secondary emission (TSE) type has been described by Batey *et al.*[88] in which the dynodes are layers of KCl supported on alumina (fig. 12.5). Gains of up to five per stage are usual with a stage potential of about 4 kV. Figure 12.6 shows the photon gain resulting from such a tube using five dynodes, while fig. 12.7 gives a similar function for a cascade type of tube.

[*Courtesy of EEV Co. Ltd.*]

Fig. 12.5. Photograph of a TSEM image intensifier.

A general comparison between the two types of tube shows that the main deficiency of the cascade tube is its slow response while its main advantage is the improved contrast possible. Both properties are due to the incorporation of intermediate phosphors[89].

Compared with a TSE membrane a phosphor/photo-cathode sandwich cannot avoid the finite build-up and decay time of the phosphor. The response time of the secondary emission process is much shorter. On the other hand the automatic contrast enhancement of the phosphor gives the cascade tube a definite advantage over one using the TSEM principle.

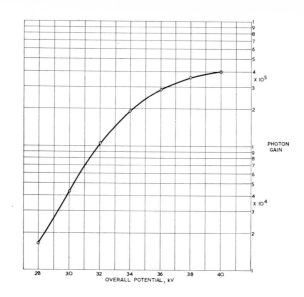

Fig. 12.6.   Typical relationship between light gain and overall potential for a TSEM tube.

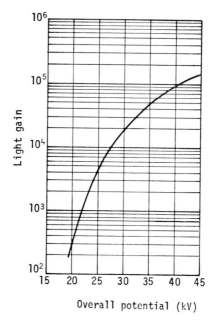

Fig. 12.7.   Typical relationship between light gain and overall potential for a cascade type tube.

124

# REFERENCES

(1) BECQUEREL: *C. r. hebd. Séanc. Acad. Sci., Paris,* **9,** 145, 1839.
(2) WILLOUGHBY-SMITH: *J. Soc. Telegraph Engrs,* **2,** 31, 1873.
(3) (*a*) HERTZ: *Annln Phys.,* **31,** 983, 1887.   (*b*) HALLWACHS: *Annln Phys.,* **33,** 307, 1888.
(4) HUGHES and DU BRIDGE: *Photoelectric Phenomena* (McGraw-Hill).
(5) BOHR: *Theory of Spectra and Atomic Constitution* (Cambridge University Press), 1922.
(6) JOHNSON: *Spectra* (Methuen Monograph).
(7) SOMMER: *Photoemissive Materials* (Wiley,) 1968.
(8) EINSTEIN: *Annln Phys.,* **17,** 132, 1905.
(9) SCHRODINGER: *Collected Papers on Wave Mechanics* (Blackie), 1939.
(10) ELSTER and GEITEL: *Annln Phys.,* **38,** 40, 1889; **41,** 161, 1890; **42,** 564, 1891.
(11) MCGEE, KHOGALI, GANSON and BAUM: *Adv. Electronics* and *Electron Phys,* A, **22,** 20, 1966.
(12) MCGEE and LUBSZYNSKI: *J. Instn Elect. Engrs,* **84,** 468, 1939.
(13) SOMMER: *RCA Rev.,* **28,** 543, 1967.
(14) SOMMER: *Proc. Phys. Soc.,* **55,** 145, 1943.
(15) SOMMER and SPICER: *J. Appl. Phys.,* **32,** 1036, 1961.
(16) SOMMER: *Optica Acta,* **7,** 121, 1960.
(17) SOMMER: *Rev. Scient. Instrum.,* **26,** 725, 1955.
(18) SOMMER: *Rev. Scient. Instrum.,* **28,** 655, 1957.
(19) English Electric Valve Company Limited process.
(20) MCCARROLL: *J. Phys. Chem. Solids,* **26,** 191, 1965.
(21) *Phototubes and Photocells* (RCA Technical Manual), Part 60, p. 77, 1963.
(22) GOETZE: *Adv. Electronics Electron Phys.,* **16,** 145, 1962.
(23) *Electronics:* p. 226, 11 November 1968.
(24) WHITEHEAD: *Principles of Television Engineering* (Iliffe). AMOS and BIRKENSHAW: *Television Engineering* (Wireless World/Iliffe).
(25) GARRATT and MUMFORD: *Proc. Instn Elect. Engrs,* **39,** 25, 1952.
(26) CAMPBELL-SWINTON: *Nature, Lond.,* **78,** 151, 1908.
(27) EUGENE LYONS: *David Sarnoff—A Biography* (Harper and Row), 1966.
(28) ZWORYKIN: *Proc. Inst. Radio Engrs,* **22,** 16, 1934.
(29) MCGEE: *Proceedings of the World Radio Convention,* Sydney (IRE) 1938.
(30) AMOS and BIRKENSHAW: *Television Engineering* (Wireless World/Iliffe), p. 196.
(31) LUBSZYNSKI and RODDA: British Patent No. 442,666, 1934.
(32) COPE, GERMANY and THEILE: *J. Br. Inst. Radio Engrs,* p. 139, 1952.
(33) LUBSZYNSKI: British Patents Nos. 468,965 (1936) and 522,458 (1938).
(34) ROSE and IAMS: *RCA Rev.,* **4,** 189, 1939.
(35) ROSE, WEIMER and LAW: *Proc. Inst. Radio Engrs,* **34,** 424, 1946.
(36) MCGEE: *Proc. Instn Elect. Engrs,* **97** (Part III), 377, 1950.
(37) TURK and MCGEE: British Patent No. 600,520, 1945.
(38) WEIMER, FORGUE and GOODRICH: *Electronics,* **23,** 70, 1950.
(39) SCHADE: *RCA Rev.,* **9,** 520, 1948.
(40) HENDRY and TURK: *J. Soc. Motion Pict. Telev. Engrs,* **69,** 88, 1960.

(41) TURK: *J. Soc. Motion Pict. Telev. Engrs*, **75**, 841, 1968.   KNIGHT: *R. Telev. Soc. J.*, **12**, 58, 1968.

(42) BEURLE: Unpublished English Electric Valve Company report.

(43) WEIMER: *RCA Rev.*, **10**, 376, 1949.

(44) HENDRY: British Patent No. 847,320, 1958.

(45) BEURLE, PAY and TURK: *R. Telev. Soc. J.*, **11**, 254, 1967.

(46) FERNSEH: British Patent No. 834,497, 1956;   Marconi British Patent No. 899,980, 1960.

(47) (a) BANKS and TURK: *I.E.E.E. Broadcast Symposium*, Washington, D.C. September 1964.   (b) BANKS: British Patent No. 1,048,390.

(48) TURK: *Sound Vision Broadc.*, **3**, 3, 1962.

(49) THEILE: *J. Telev. Soc.*, **9**, 1, 1959.

(50) RYE: *Electron. Equip. News*, May 1967.

(51) SCHADE: *RCA Rev.*, **9**, 29, 1948.

(52) ROBINSON: *RCA Tube Tips*, No. 138, 1962.

(53) WEAVER: *BBC Engng Monogr.*, No. 24, 1959.

(54) EDWARDSON: *BBC Engng Monogr.*, No. 37, 1961.

(55) HOLDER: *International Television Conference Report* (Institution of Electrical Engineers), p. 201, 1962.

(56) GROSSKOPF: Private Communication.

(57) NEUHAUSER: *J. Soc. Motion Pict. Telev. Engrs*, **68**, 455, 1959.

(58) THEILE and PILZ: *Arch. elekt. Übertr.*, **11**, 17, 1957.

(59) WEIMER: *RCA Rev.*, **10**, 366, 1949.

(60) VAN ASSELT: *Proc. Natn. Electron. Conf.*, U.S.A., 1968.

(61) KLEM: Fernseh Technische Gesellschaft Meeting, October 1968, Saarbrucken.

(62) FORGUE, GOODRICH and COPE: *RCA Rev.*, **12**, 335, 1951.

(63) (a) DE HAAN, VAN DER DRIFT and SCHAMPERS: *Philips Tech. Rev.*, **25**, 133, 1963/64.   (b) VAN DOORN, *Philips Tech. Rev.*, **2 7**, 1, 1966.   (c) VAN ROOSMALEN, *Philips Tech. Rev.*, **28**, 60, 1967.

(64) GORDON: *Bell Labs Rec.*, p. 175, June 1967.

(65) CROWELL, BUCK, LABUDA, DALTON and WALSH, *Bell Syst. Tech. J.*, p. 491, February 1967.

(66) LUBSZYNSKI: British Patent No. 827,058, 1955.

(67) WEIMER: *Adv. Electronics Electron Phys.*, **13**, 387, 1960.

(68) LUBSZYNSKI, TAYLOR and WARDLEY: *J. Br. Inst. Radio Engrs*, May 1960.

(69) (a) HEIJNE: *Acta electron.*, **2**, 124, 1957.   (b) HEIJNE, SCHAGEN and BRUINING, *Philips Tech. Rev.*, **16**, 23, 1954.

(70) BURNS and DAVIS (EMI): *Optical Effects in Photocathodes*, Private Communication from the Authors.

(71) VINE, NOVICE: *Appl. Optics*, **6**, 1171, 1967.   See also *Proceedings of the Fourth Symposium on Photoelectric Image Devices* (to be published).

(72) MORTON and RUEDY: *Adv. Electronics Electron Phys.*, **12**, 183, 1960.

(73) (a) Philips Gloeilampenfabrieken: British Patent No. 326,200, 1928.   (b) McGEE, AIREY, ASLAM, POWELL and CATCHPOLE: *Adv. Electronics Electron Phys.* A, **22**, 113, 1966.

(74) SCHNEEBURGER, SKORINKO, DOUGHTY and FEIBELMAN: *Adv. Electronics Electron Phys.*, **16**, 235, 1962.

(75) GUILLARD and CHARLES: *Adv. Electronics Electron Phys.* A, **22**, 315, 1966.

(76) McGEE, WILCOCK and MANDEL: *Adv. Electronics Electron Phys.*, **16**, 145, 1962.

(77) GOETZE: *Adv. Electronics Electron Phys.* A, **22**, 219, 1966.

(78) SKAGGS: Texas Instruments Inc. Report.

(79) HORTON, MAZZA and DYM: *Proc. Inst. Elect. Electron. Engrs*, **52**, 1513, 1964.

(80) SANDBANK: *Solid-St. Electron.*, **10**, 369, 1967.

(81) Plessey Company's Integral Array SC12.

(82) WEIMER *et al.*: *Proc. Inst. Elect. Electron. Engrs,* **55,** 1591, 1967.
(83) De Oude Delft Publication from NV Optische Industrie, De Oude Delft.
(84) SMYTH, POYNTON and SAYERS, *Proc. Instn Elect. Engrs,* **110,** 16, 1963.
(85) MULLARD LIMITED: *Screen Phosphors and Industrial Cathode Ray Tubes,* 1964.
(86) MCGEE, KHOGALI and GANSON: *Adv. Electronics Electron Phys.* A, **22,** 11, 1966.
(87) MCGEE, ARRAY, ASLAM, POWELL and CATCHPOLE: *Adv. Electronics Electron Phys.* A, **22,** 113, 1966.
(88) BATEY and SLARK: *Adv. Electronics Electron Phys.* A, **22,** 63, 1966.    SLARK and WOOLGAR: *Adv. Electronics Electron Phys.,* **16,** 141, 1962.
(89) EMBERSON: *Adv. Electronics Electron Phys.* A, **22,** 129, 1966.

# FREQUENCY CONVERSION

## Masers and Lasers    Part III

### M. J. Beesley
### Services Electronics Research Laboratory

*Acknowledgement*

The author wishes to thank Miss H. Shearman and Mr. A. G. Wilson for the assistance so freely given with the manuscript, and to the Ministry of Defence (Navy Dept.) for permission to publish.

# INTRODUCTION

THE conversion of one frequency into another is exemplified in a most spectacular form by the relatively recent development of the laser. This powerful source of radiation has acted as a considerable stimulus to the field of optics and many interesting applications seem probable. Although frequency conversion is the basic mechanism of the laser, it is not the change in frequency as such which is so significant, but rather that there is usually a fundamental difference between the character of the radiation which is emitted from the device and the energy which is supplied to it. Before describing how the construction of the first masers and lasers came about, we shall discuss the characteristics of their output, as an understanding of these is necessary for the significance of the laser to be appreciated.

# CHAPTER 3.1

## coherent sources

### 1. *Coherence*

THE beam of light emitted from a laser is said to be almost coherent. Conventional sources, however, such as an electric light bulb, a sodium lamp, a fluorescent tube or the sun, are said to be incoherent. In theory it is possible to take a mercury or cadmium lamp and make it more coherent than most lasers; however, the intensity of the light produced is then many orders of magnitude less than that from a laser, so it can be said that for all practical purposes only the laser can give a powerful coherent beam of light. It is this previously unobtainable characteristic that is the outstanding feature of the laser and makes it such an important discovery in modern physics.

The light produced by a laser can be thought of as a wave oscillating some $10^{14}$ times a second and of wavelength approximately $1/100$ mm; for such a wave to be coherent two conditions must be fulfilled—first it must be of very nearly a single frequency, that is the spread in frequency or line-width must be very small. Secondly, the wave-front, the surface formed by all points of equal phase, must have a constant shape not varying from one moment in time to another. The wavefront of the light emanating from a point source, for instance, will be spherical and that of a parallel beam of light will be plane. To obtain a better understanding of these ideas it is necessary to consider the nature of light itself.

In the opening years of this century, certain curious phenomena were observed by physicists which could only be accounted for satisfactorily by describing radiation in terms of particles (called photons) rather than waves. For the previous hundred years the wave theory of light was accepted universally, as it explained so well such effects as interference, polarization and diffraction, and in consequence the corpuscular theory of Newton had been discarded. Nevertheless, with the advent of Einstein's photon theory in 1905, physicists had no alternative but to accept that waves and particles were both valid descriptions of the same thing. The idea of the photon came five years after the famous work of Max Planck who had shown that any change in the energy of a radiation field must take place in discrete steps. It will be shown later how this work of Planck on black-body radiation was to have an important bearing on the very beginnings of the laser. Earlier work on line spectra, the discovery of the electron by J. J. Thomson in 1897 and the subsequent work of Rutherford paved the way for Niels Bohr to unite all these ideas in his atomic model of 1913. According to Bohr, electrons rotate in orbits about an atomic nucleus just as the planets rotate about the sun. Only discrete orbits are allowed, however, and when an electron changes its orbit from one of higher energy to one of lower energy, radiation is emitted in the form of a photon having an energy $E = h\nu$, where $\nu$ is the frequency of the wave associated with the photon, $E$ is the difference in the energies of the two orbits and $h$ is a constant called Planck's constant.

Having outlined the process by which photons of radiation are emitted we can now relate such emission to the concept of coherence. Suppose an electron changes its orbit around an atom to a new one of lower energy, then a photon having discrete energy $E$ is emitted. The emission of the photon (or the change in energy of the atom) can be said to take place in a finite time which is denoted by $\Delta t$ and called the lifetime. For atoms in free space, i.e. isolated from external influences, $\Delta t$ has a value of about $10^{-8}$ s, and so the photon can be considered as a wave of finite length; the wave having a frequency $\nu = E/h$ and an amplitude which rises initially from zero to a peak and then drops again to zero when the electron has completed its descent to the lower energy level. Such a wave is known as a Gaussian wave train and its shape is shown in fig. 1.1. The meaning of $\Delta\nu$ is explained below. Usually waves of many different frequencies are emitted corresponding to different energy changes within the atom, with the result that the spectrum consists of a collection of single lines; however, as far as this discussion is concerned, we are only considering just one line arising from one particular transition.

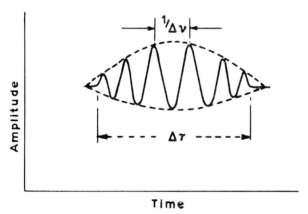

Fig. 1.1. Gaussian wave-train.

The light which is emitted from a gas discharge lamp, for example, will consist of a very large number of such wave trains each wave train being radiated from those atoms in the gas whose electrons are changing their orbits, or as is conventionally said, undergoing transitions from higher to lower energy states. The energy lost by the lamp in this way is derived from the electrical power supplied to it. Now suppose for the sake of argument that it is possible, with some imaginary instrument, to watch some chosen point in the path of the wave-trains emitted by the lamp and that our instrument can detect changes in amplitude and phase over very small periods of time (better say than $10^{-15}$ s. which is roughly the time taken for one cycle of the wave to pass the point of observation). It is then possible, as each Gaussian wave train passes the observation point, to watch the amplitude of the wave rise and fall. If the frequency is $10^{14}$ Hz and $\Delta t$ is $10^{-8}$ s then about a million undulations in amplitude will be observed. As the atoms in the gas radiate in a random fashion, it is impossible to predict when the next wave-train will arrive at the point of observation after the previous train has passed by; although once the front end of a wave-train reaches the observation point a prediction can be made about the amplitude and phase at some later time, assuming of course that the wave-train is still passing the observation point at the later time. It is this ability to predict amplitude and phase that is the essence of coherence and the light is said to be coherent for the time, $\Delta t$, which the wave-train takes to pass the point of observation, the longer this time, the greater the coherence. For obvious reasons, the lifetime $\Delta t$ is also known as the coherence time and the type of coherence described above is known as time coherence. The actual length of the wave-train, $L$, which is simply obtained by multiplying the coherence time by the velocity of light $c$, is known as the coherence length and is therefore given by:

$$L = c\Delta t. \tag{1.1}$$

Further insight can be gained into the concept of time coherence by applying a mathematical technique known as Fourier analysis to the Gaussian wave-train. We have seen that the radiation emitted by an atom, as a consequence of an electronic transition between two energy levels, consists of wave-trains of finite length and single frequency. If, however, Fourier analysis is applied to one such wave-train it can be shown that it is equivalent to a number of infinitely long wave-trains of differing frequencies spread about a central frequency $\nu$. The spread, or linewidth, of these frequencies is measured at the half amplitude points and is denoted by $\Delta \nu$. The shape of a line is indicated in fig. 1.2. This is often a more convenient description of the finite wave-train but it must be emphasized that both descriptions are exactly equivalent.

It is useful to obtain a relation between $\Delta \nu$, the linewidth of the light source, and the coherence length $L$. To do this we make use of another result of Fourier analysis which states that

$$\Delta \nu . \Delta t \sim 1. \tag{1.2}$$

From this relation it can be seen that a source of light which generates long wave-trains and which is said to be more coherent than one

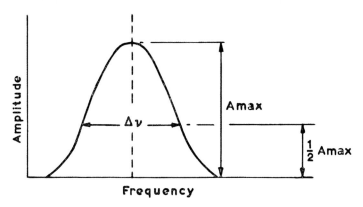

Fig. 1.2. Gaussian wave-train.

generating shorter trains, and also has a larger value of $\Delta t$, must hence have a smaller value of $\Delta \nu$. Long time coherence is therefore equivalent to a narrow line-width. An expression of time coherence can therefore take three forms:

(a) the coherence time, which is the time taken for the wave-train to pass the point of observation;

or

(b) the frequency line-width $\Delta \nu$ or the wavelength linewidth $\Delta \lambda$;

or

(c) the length of the wave-train or coherence length, $L$.

The relation between coherence length and linewidth is derived as follows:

we have from equation (1.1):

$$L = c \Delta t,$$

hence from equation (1.2):

$$L = c / \Delta \nu, \tag{1.3}$$

but

$$c = \nu \lambda; \tag{1.4}$$

differentiating (1.4) we have:

$$\Delta \nu = \frac{c}{\lambda^2} \Delta \lambda \tag{1.5}$$

(ignoring the negative sign), and so substituting (1.3) in (1.5) we finally obtain:

$$L = \frac{\lambda^2}{\Delta \lambda}. \tag{1.6}$$

135

We can now calculate the coherence length of a wave-train resulting from a typical free space atomic energy level transition having a lifetime of $10^{-8}$ s. We have:

$$\Delta t \sim 10^{-8} \text{ s};$$

$$\therefore \quad \Delta \nu \sim 10^8 \text{ Hz} \quad \text{from equation } (1.2).$$

If we assume green light for which

$$\lambda = 0.5 \times 10^{-4} \text{ cm},$$

then from (1.5)

$$\Delta \lambda = 0.001 \text{ Å},$$

from which finally

$$L \sim 3 \text{ metres}.$$

This result applies to the ideal case of a single atom; when a collection of such atoms are in close proximity, as is the case, for example, of a gas enclosed in a container, then the natural linewidth, as the linewidth corresponding to the former situation is called, has to be replaced by a larger linewidth. This broadening of the linewidth has two principal causes: the presence of other radiating atoms (pressure broadening) and the motion of radiating atoms (Doppler broadening). Despite these effects a low pressure cadmium lamp can give a line whose width is only about 0.01 Å with an approximate coherence length of 30 cm. Most normal sources of light, however, consist of a large number of much broader lines each having a coherence length which is very small indeed —perhaps a fraction of a millimetre.

In the case of lasers, the situation is quite different; this is because the photons emitted do not result simply from random transitions between energy levels of atoms but by a chain reaction process whereby one photon stimulates a downward transition and so produces two photons which represent two waves *in phase*. The result of very many of these processes is to make the assembly of atoms enclosed within the laser act as an amplifying medium. This process and its effects will be discussed in detail later. We have seen that the time for emission of a photon (the lifetime) which is equivalent to the coherence time, is about $10^{-8}$ s. for an isolated atom. In the case of a photon emitted by a laser this time is increased by virtue of the amplification which takes place between the mirrors that form the resonating part of the laser to a value of about $10^3$ s, and so the line-width, from equation (1.2) is $10^{-3}$ Hz or $10^{-14}$ Å. This fantastically small width is never observed in practice as over a period of time the actual frequency of the line $\nu$ changes its value slightly due to disturbances to the laser resonator which are caused by such things as vibration, air currents and temperature fluctuations. Nevertheless, line-widths as small as 20 Hz have been observed—the corresponding coherence length is of the order of 15 000 km. For reasons, which will become apparent later, unless special precautions are taken, most lasers give an output which consists of a number of narrow lines each separated by typically 150 MHz, so an approximate value of the overall linewidth of the output may be 1000 MHz, resulting in a coherence length of about 30 cm. Now we have seen that this sort of coherence length is available from conventional light sources, albeit of a special form such as a low pressure cadmium lamp; why then is the laser such a revolutionary source? The answer to this question is that the laser gives a narrow beam of light, whose wave-front can be very nearly plane, *and of very high intensity*.

In addition to time coherence, which is associated with the linewidth of the source, there is another aspect of coherence which is termed space coherence. For a source to be completely coherent it must have spatial as well as temporal coherence. It has been explained that time coherence implies the possibility of predicting the phase at some later time if the phase at some initial time is known and it must therefore follow that if the time interval between the initial and later time is kept constant, then a number of observations and predictions about phase and amplitude can be made by simply choosing different initial times. If the source is time coherent then the phase difference in each case will be the same and the time interval over which the phase difference remains constant is the coherence time. In the case of spatial coherence, however, we are not concerned with different observations at different points, in time, along the wave-train but at different points, in space, on the wave-front. The wave is said to be spatially coherent if there is a constant phase difference between any two chosen points on the wave-front. By 'constant' a time much longer than one cycle or $10^{-15}$ s is meant. If this is not so, then both a spatially and temporally incoherent source would appear coherent for such a very short observation period. This happens because any change in phase would not be apparent anyway within a period which is a fraction of a cycle.

Time and space coherence are really two aspects of the same phenomenon. When we speak of time coherence we are saying that the relative phases between two points *in time* must remain constant over some time interval; space coherence, on the other hand, involves the relative phases between two points *in space* remaining constant again over some time interval. In each case, the longer the time interval, the greater the coherence.

In order to clarify these ideas in a practical way it is helpful to consider two experiments, one performed by Michelson and the other by Young. In the Michelson interferometer, shown in fig. 1.3, a wave-front is produced by the source S and is split by the beam-splitter B to provide two wave-fronts of identical shape and intensity. The two mirrors $M_1$ and $M_2$ serve to recombine the two wave-fronts at the observation point O, which is usually a telescope.

Consider a wave-train emitted from S. Then if the distance between the beam-splitter and each mirror is the same (the beam-splitter being

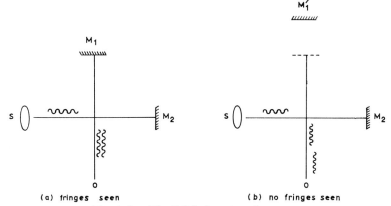

(a) fringes seen          (b) no fringes seen

Fig. 1.3. The Michelson interferometer.

assumed infinitely thin), the two wave-trains produced by the beam-splitter will overlap exactly as they travel from the beam-splitter to the observer because the path lengths they have each traversed are equal and as a result interference fringes will be seen on looking through the microscope. If, however, one of the mirrors is moved so that the path traversed by one wave-train is longer (or shorter) than that traversed by the other by at least its own length (as is indicated by M$_1'$ in fig. 1.3 b), interference fringes will not be seen because the two trains will not exactly overlap when they reach O. If the mirror M$_1$ is moved perpendicular to its plane so that fringes are just not seen, then the distance through which the mirror has been moved is about half the length of the wave-train. In this way the length of the wave-train, i.e. the coherence length, is easily measured, and from this can be deduced the coherence time—the latter being the time taken for the wave-train to traverse the greatest possible extra path length consistent with fringes still being visible. In the Michelson interferometer spatial coherence is not so important because, although the phase of S may fluctuate across its wave-front, the *relative* phase difference between the two recombined beams at O will remain constant.

Space coherence is important in a situation like Young's two-slit experiment shown in fig. 1.4. Here two sources of light are emitting waves which are combined to form fringes on a screen.

With a conventional light source it is necessary to place a small hole H$_1$ as indicated so that H$_2$ and H$_3$ are effectively illuminated by the same point source, i.e. to ensure spatial coherence between them. H$_2$ and H$_3$ then act as two secondary sources emitting waves to form interference fringes on the screen. Suppose, instead, that a long fluorescent light tube had been used as a light source as shown in fig. 1.5. In this case no fringes will be seen on the screen as there is no spatial coherence between the portion of the tube supplying light to H$_1$ and that supplying

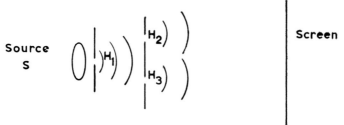

Fig. 1.4. Young's two-slit experiment.

light to $H_2$. If it were possble to take a snapshot of the screen in say $10^{-15}$ s, then a fringe system would of course be recorded, but another snapshot of similar duration taken at a later time would show fringes in different positions due to the change in relative phase. Thus, over a 'long' period of time the fringes average out, as far as the human eye is concerned, and an even level of illumination is observed. Now suppose that the fluorescent tube is replaced by a laser as shown in fig. 1.6. In this situation fringes are seen because the laser is spatially coherent. The laser is depicted as giving a plane wave but this is not necessary; only consistency of phase difference between $H_1$ and $H_2$ is required.

Fig. 1.5. Young's two-slit experiment using a fluorescent tube as the source.

If, in fig. 16, a ground glass diffuser were placed between the plane wave from the laser and the two slits then, in general, fringes would still be seen on the screen. They would not be seen, however, if the piece of ground glass was rapidly moved to and fro parallel to the plane of the slits because constant phase difference between the two slits over a period of time would not occur, i.e. the light reaching the slits would be spatially incoherent.

This ability of coherent light, and hence of lasers, to form interference fringes with great facility is of considerable value in many applications. The recent up-surge of work on holography, for instance, by which three-dimensional pictures of objects are formed, is almost entirely due to lasers becoming available.

139

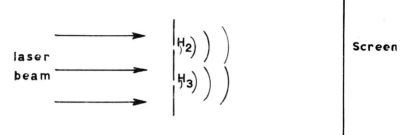

Fig. 1.6. Young's two-slit experiment using a laser beam as the source.

Laser light is always spatially coherent and in certain circumstances, not only is the shape of the wave-front constant, but it is also almost flat. Thus the output from a laser appears as a narrow beam which diverges only very gradually. In this situation the laser is said to be operating in a *uniphase* mode and in this state the beam can be focused, by means of a lens, down to a point which is almost as small as it is theoretically possible to achieve, a theoretical limit being due to the finite wavelength of light. Such a point is called a diffraction limited point and immense energy densities can be obtained at such point suitable for burning or welding.

Figure 1.7 shows how such a uniphase temporally coherent wave can be obtained from an ordinary source by taking a low-pressure cadmium lamp and placing a small pinhole in front of it. If a lens is then placed such that its focus is at the pinhole then a plane (uniphase) wave-front is produced. However, it must be remembered that such a wave-front would be of extremely small intensity—many orders of magnitude less than that produced by a laser.

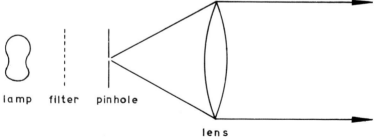

Fig. 1.7. Production of highly coherent light from a conventional source.

To summarize, therefore, the laser, unlike other light sources, is capable of producing uniphase, almost single frequency, outputs of extremely high intensity.

Having briefly described the nature of laser radiation, we shall now proceed to a discussion of how it is obtained.

## 2. *Stimulated emission*

According to Bohr's theory, an atom can be thought of as a collection of electrons rotating in orbits around a central nucleus. Each orbit of an electron is associated with a certain energy and so an electron changing its orbit causes the atom to change its energy; if an electron moves to an orbit of higher energy, the atom gains energy and if it changes to an orbit of lower energy the atom loses energy. It is thus possible to describe any atomic or molecular system by a series of energy levels, each level representing a possible energy state of the atom or molecule. We shall now describe the three processes by which an electron can change energy levels.

If energy in some form is supplied to an atom, the emission of light may occur. This is explained by assuming that the supplied energy excites an electron from some lower energy state to a higher energy state. At some later time the electron falls back to its original energy state and energy is emitted in the form of a photon of frequency proportional to the energy difference between the two levels. The emission of a photon in this way takes place at some unpredictable time after excitation and hence, because of its random nature, is known as spontaneous emission.

The energy used to excite an electron from a lower state to a higher state may be in the form of a photon having energy equal to the energy difference between the two levels. In this case the photon is absorbed by the atom and consequently this process is known as absorption.

Although spontaneous emission is a random process it is possible to specify an average length of time before emission occurs; this is the lifetime introduced previously and is of the order of $10^{-8}$ s for an atom in free space. It should be realized that, in the case of a gas at a relatively high pressure, collision of an excited atom with another atom may take place before spontaneous emission occurs so that energy is lost by collision rather than by the emission of a photon and consequently the gas will heat up rather than emit light. At lower pressures emission will occur before collision. If the electron returns to its original state, then the photon emitted will be of the same frequency as that absorbed—the emitted radiation then being called resonance radiation. The excited electron can also return to its original state by means of intermediate levels and so a number of photons of lower energy are produced, this process being known as fluorescence. A good example of this type of frequency conversion is the excitation of iondine vapour by green light with the subsequent observation of red fluorescent light (of lower frequency). Absorption is illustrated in a notable way by the existence of dark lines in the spectrum of the sun. These lines correspond to the positions of the lines characteristic of elements observed in the laboratory and so the composition of the sun can be inferred. The sun has a very hot, dense central portion called the photosphere and the radiation spectrum from this region is continuous on account of the high temperature and very close proximity of the atoms which result in energy levels being disturbed so much that a very large number are possible which are so close together as to make the spectrum appear continuous. Some of this radiation is absorbed by the cooler outer region called the chromosphere—the absorption thus being characteristic of the elements making up the chromosphere and so causing dark lines, where absorption has taken place, to be observed in the spectrum. These lines are known as Fraunhofer lines, after their discoverer who first noted their existence in 1815. It is interesting to note that the helium lines were observed by Lockyer in 1868 before the discovery of helium on earth.

Figure 1.8 shows diagrammatically the two processes; spontaneous emission and absorption, which have been described so far.

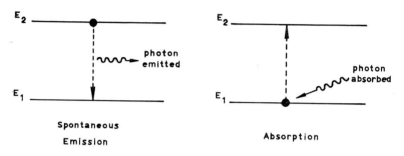

Fig. 1.8. Spontaneous emission and absorption.

Until 1917 spontaneous emission and absorption accounted for the physical phenomena observed, but when Einstein attempted to analyse the thermal equilibrium of an atomic system that was absorbing and emitting radiation, he found that it was necessary to postulate another process whereby a photon of energy equal to the difference in energy between two states interacts with an electron in the upper energy state of a two-level system as shown in fig. 1.9. In this case the electron does not descend to the lower level of its own accord, as in spontaneous emission, but is forced, or stimulated down by the arrival of the photon; this therefore produces an additional photon and the system has therefore changed from one photon and an excited electron to two photons and an unexcited electron. The two photons are not only of the same energy but also represent waves *which are in phase*. This third process is called, for obvious reasons, stimulated emission and is the basis of laser operation.

Once a stimulated emission takes place within a collection of atoms then all subsequent stimulated emissions which occur as an indirect result of the first will form wave-trains which are all in phase and hence coherent. This is, therefore, quite different from the situation in spontaneous emission where one such emission is quite independent of any other and so the resulting radiation must be incoherent. Stimulated emissions can be considered as a type of chain reaction, the original

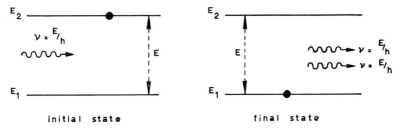

Fig. 1.9.   Stimulated emission.

photon being 'amplified' by subsequent stimulated emissions.   The
meaning of the acronym LASER (*L*ight *A*mplification by the *S*timulated
*E*mission of *R*adiation) should now be obvious.   Three stages of this
amplification process are depicted in fig. 1.10.

Stimulated emission is not observed in the visible region of the
spectrum under normal laboratory conditions and for many years after
Einstein's work it remained a theoretical curiosity.   In the next section
we shall see why this was so.

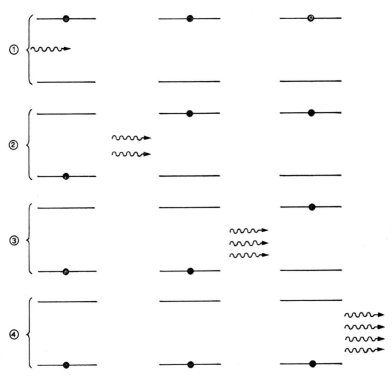

Fig. 1.10.   The amplification process.

143

## 3. Population inversion

Consider two two-level energy systems representing an atom in an upper and lower energy state. Suppose that a photon of energy equal to the energy difference between the two levels approaches the two systems; then which event is most likely to take place, absorption or stimulated emission? The answer is that, under normal circumstances, both processes are equally probable (Einstein's proof of this will be given in the next chapter). It is thus apparent that, in a system consisting of a very large number of such atoms (or molecules), the dominant process will depend on the number of atoms which are in the upper and lower states. A larger population (i.e. number of atoms) in the upper level will result in stimulated emission dominating, while if there are more atoms in the lower level, absorption will be the dominant process. Under conditions of thermal equilibrium, the populations of a number of energy levels obey what is known as a Boltzmann distribution, which means that in a two-level system the population of the upper state will be much smaller than that of the lower state; to be precise $1/e$ of that of the lower state if the energy difference is $kT$ ($k$ is a constant called Boltzmann's constant and $T$ is the absolute temperature). In terms of electron volts, which are convenient units of energy, $kT$ at room temperature is about 0·025 eV. The first energy level above the lowest possible, which is called the ground state, is, for most atoms and molecules, separated by a gap of about 1·25 eV. This amount of energy corresponds to that of a photon of visible light, i.e. a photon emitted as a result of a transition from the first excited state to the ground state would be equivalent to a wave-train of wavelength approximately 5000 Å. Therefore, under normal equilibrium conditions and at room temperatures, the population of the first excited state will be negligible and hence the amount of stimulated emission will also be negligible. This is the reason why, at visible wavelengths, stimulated emission is not usually observed. For stimulated emission to dominate it is necessary to increase the population of the upper energy level so that it is greater than that of the lower energy level—this situation is known as *population inversion*, and is an essential condition for any laser to operate.

## 4. The ammonia maser

Although the basic principles of stimulated emission and hence the fundamentals of the laser were established early in the century, it was to be many years before the first device employing these principles was built. This came about through the necessity to develop radar sets employing shorter and shorter wavelengths. By this time the various pieces of hardware that were needed had also become available—an example of technology having to catch up with theory. During the second world war Charles H. Townes was engaged on radar work and was endeavouring to develop airborne radar sets of very short wavelengths in order to overcome jamming. Townes was dubious about the feasibility of using microwaves of wavelengths shorter than a few centimetres because such waves suffer strong absorption by atmospheric water vapour. Nevertheless 1·25 cm sets were built and, as expected, were found to be somewhat unsatisfactory. However, in investigating these waves, Townes found that they were strongly absorbed by ammonia vapour; it was this observation that was to provide the clue to the operation of the first laser. The first laser was to operate by means of ammonia molecules at microwave frequencies and was called a maser, the ' m ' standing for ' microwave ' or ' molecular '. The word laser was coined later and now seems to be applied to any wavelength shorter than the far infra-red.

In 1951 the idea occurred to Townes that the ammonia molecule itself might act as a source of microwaves. His earlier observations had indicated a strong absorption, i.e. a transition, at microwave frequencies and if a sufficient number of ammonia molecules could be excited to a higher state, then the introduction of a weak signal of appropriate frequency would result in an amplified output.

The ammonia molecule is depicted in fig. 1.11. It consists of two forms, normal and inserted, the normal form being of lower energy than the inverted form. The difference in energy between these inversion forms, as they are called, corresponds to a frequency of 23 870 MHz or a wavelength of about 1·25 cm. The ammonia maser operates simply as a result of a transition between an upper and lower level, no other levels

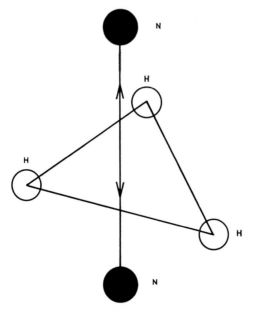

Fig. 1.11. The ammonia molecule.

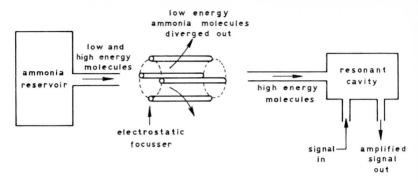

Fig. 1.12. The ammonia maser.

being involved, and is therefore called a two-level maser. A diagram depicting the essential components of the ammonia maser is shown in fig. 1.12.

A directed supply of ammonia molecules is obtained by connecting a series of fine tubes to a tank containing ammonia at a pressure of a few torr. The beam of molecules is then passed through an electrostatic focuser which has the effect of converging those molecules in the high energy state and diverging those in the low energy state. The focuser itself consists of a cage of four rods 55 cm long and 1 cm apart which act as electrodes. Two opposite electrodes are earthed while the other two are kept at a potential of 15 kV. By this ingenious method of simply discarding molecules in the lower state, population inversion is achieved. The focused beam of high energy molecules is then passed into a cavity designed to be resonant at a frequency of 23 870 MHz. Thus, if a microwave signal of this frequency is fed into the cavity, it can only stimulate downward transitions and is hence amplified. If enough high energy molecules are injected into the cavity then a spontaneous emission can start a self-sustaining chain reaction of stimulated emissions and the maser will then act as an oscillator.

The output from an ammonia maser oscillator is of very low power (about $10^{-10}$ watt) but of very high spectral purity (a line-width of 5 Hz, or one part in $10^{10}$, over a period of a minute is possible). As an amplifier, it is limited by its narrow amplification band-width of only a few kilohertz about the central frequency and, of course, no tuning is possible. For these reasons the most important application of the ammonia maser is as a frequency standard.

5. *The three-level maser*

The search for a tunable maser amplifier of higher power and wider band-width led Bloembergen to propose the idea of the three-level maser in 1956. Two years later, in 1958, Feher, Scovil and Seidal of the Bell

146

Telephone Laboratories successfully constructed such a maser using a crystal of gadolinium ethyl sulphate as the active medium.

It has been indicated that a necessary condition for laser action to take place is the establishment of a population inversion between two levels of an atomic or molecular system. In the case of the ammonia maser this is achieved by the physical removal of molecules in the lower energy state from the system. Generally speaking, such a technique is not possible and a more subtle way round the problem must be found. Bloembergen's scheme was to select a suitable three-level system in which, because of the Boltzmann distribution, the population of each energy level would decrease from the bottom level to the top level and then to 'pump' atoms from the bottom level to the top by supplying photons of the correct frequency. As absorption is as equally possible as stimulated emission, the system would stabilize with the populations of the bottom and top levels equal. However, in this situation, if the relative spacing between the three levels is carefully chosen, then the population of the middle state can be made to exceed the bottom, i.e. a population inversion between the middle and bottom levels exists and so laser action is possible. Unlike the case where visible light is concerned, the energy of a microwave 'photon' is relatively very small in comparison with the thermal energy and hence the three energy levels will be relatively very close together and consequently very close to the ground state. It follows, therefore, that at room temperatures their respective populations will be almost equal. To increase the discrepancies in population the active medium is cooled to liquid helium temperatures $(4\,^\circ\text{K})$ thus ensuring a large difference in population between the levels. Figure 1.13 $a$ shows the situation before cooling,

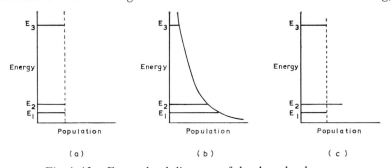

Fig. 1.13. Energy level diagrams of the three-level maser.

fig. 1.13 $b$ that after cooling and fig. 1.13 $c$ shows the situation after both cooling and pumping. It can be seen from the diagrams that the pumping energy must be $E_3 - E_1$, and that the maser output will consist of photons of energy $E_2 - E_1$. Pumping energy of frequency $E_3 - E_1/h$ has therefore been converted to energy of frequency $E_2 - E_1/h$. It should be apparent that any form of pumping in a two-level system will be unsuccessful as equality of population is the most that can be achieved unless very special procedures are adopted which are beyond the scope of this book. Any laser or maser will thus be an example of frequency conversion—the pumping 'frequency' (e.g. visible light, radio waves, direct current discharges) is always different to the laser frequency.

In order to make such a system work it is necessary to find a material which has three energy levels separated by small energy gaps equivalent to microwave frequencies. Bloembergen thought that paramagnetic substances might be ideal for this purpose. These materials are crystals whose atoms or molecules are in effect permanent magnets and under the rules of the quantum theory each atom or magnet can take up only discrete orientations with respect to an applied magnetic field. When the magnet points directly against the field it has the highest possible energy and when it points directly with the field it has the lowest possible energy. One of the most successful paramagnetic materials to be used has been ruby which consists of aluminium oxide ($Al_2O_3$) to which has been added between 0·01 and 1% of chromium, the atoms of which take up positions as ions ($Cr^{3+}$) within the crystal lattice. For a 1% dilution there are about $10^{20}$ ions cm$^{-3}$. Each chromium ion acts as a magnet or is said to have a permanent magnetic dipole moment. In the case of ruby, four different orientations of the chromium ion magnetic dipoles are possible and so it follows that four different energy levels are possible, the magnitudes of which are dependent on the strength of the applied magnetic field. By changing the field, therefore, a ruby maser can be tuned over a frequency range. Figure 1.14 shows a typical plot of energy against magnetic field strength for a given orientation of the ruby crystal with respect to the applied magnetic field.

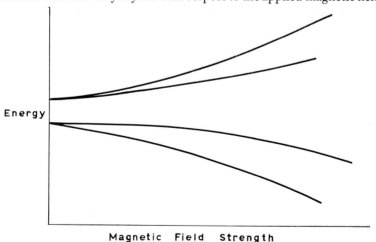

Fig. 1.14. Variation of energy levels with applied magnet field strength for ruby.

148

It is assumed, of course, that being a three-level process, one level will not be utilized.

There are many possible paramagnetic materials which can be used as maser materials but all work on the same basic principles outlined above. The first three-level maser used was not ruby, but gadolinium ethyl sulphate, in which maser action took place between energy levels of gadolinium ions, and was actually an oscillator. McWhorter and Meyer are credited with the first three-level maser amplifier which employed chromium in a host material of potassium cobalticyanide.

## 6. *The ruby laser*

In 1958 Schawlow and Townes put forward their proposals for extending the range of maser action into the visible. They suggested that the resonant cavity necessary for oscillation might take the form of a Fabry–Perot interferometer; this consists essentially of two plane parallel mirrors placed some distance apart and facing each other. In 1960 Maiman of the Hughes Research Laboratory obtained laser oscillation at 6943 Å in the red region of the spectrum using a ruby crystal as the active material. The Fabry–Perot resonator was made by simply polishing the ends of a ruby rod flat and parallel to a minute of arc and then silvering both ends, making one end almost totally reflecting. The other end was partially silvered to allow about 10% transmission and so formed the output end of the oscillator.

The ruby laser of Maiman is an example of a three-level system (see fig. 1.15) operating on a somewhat different basis to the ruby maser. It should be noted that the energy levels of interest in the ruby laser are quite different, and much further apart, than those in the ruby maser. It is simply a fortuitous coincidence that ruby is so well suited to both maser and laser operation. A very simple account of the three-level ruby laser is given now, while a more detailed description will be found in chapter three.

Chromium ions in the ground state are excited, by an intense flash of white light from a lamp, from the ground state to an upper state which actually consists of a large number of levels forming a band. This makes the pumping more efficient because an exact pumping frequency is not required. Unlike the ruby maser case the excited atoms then drop from the band of upper states to a middle state. Figure 1.15 indicates the steps in this process.

The transition down into the middle state is accompanied, not by the emission of a photon, but by the direct transfer of energy to the surrounding crystal lattice which causes the ruby rod to heat up. This latter process is known as a non-radiative transition. Once an atom reaches the middle state it spends an unusually long time there before droppping down, by spontaneous emission, to the ground state. States such as this are said to be metastable and it is because of this characteristic that the population of the middle state builds up while that of the

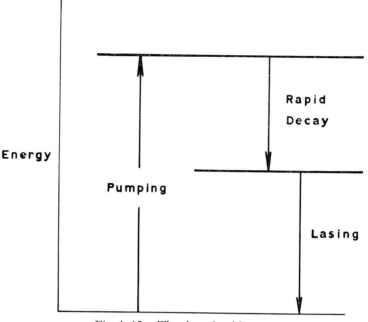

Fig. 1.15. The three-level laser.

bottom, or ground state, is depleted, i.e. a population inversion is achieved. This method of obtaining a population inversion is very inefficient because as the middle state is effectively empty at the start of the pumping process (owing to the Boltzmann thermal distribution) at least half the population of the ground level must be pumped up to the middle level before a population inversion is achieved. In addition very

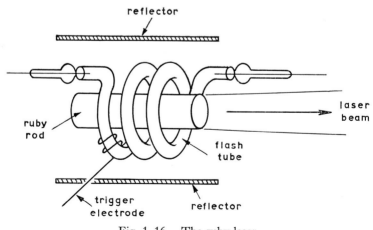

Fig. 1.16. The ruby laser.

little of the electrical energy which is supplied to the flash-lamp ends up in pumping photons and carefully designed reflectors are essential in order to concentrate as much light as possible onto the ruby rod. The pumping flash is necessarily brief and hence only pulsed operation is possible normally. A pulse duration of about 1 ms is usual with a power of as much as 10 kW.

If the laser is liquid cooled, and the light from special pumping lamps of very high intensity is focused on a small ruby crystal then a continuous output is obtained. Liquid nitrogen was used by Nelson and Boyle as the coolant when they obtained continuous operation for the first time in 1962 in the visible region of the spectrum.

The essential components of the ruby laser are shown in fig. 1.16.

### 7. *The helium–neon laser*

The first continuously operating laser was constructed by Javan, Bennet and Herriot at the Bell Telephone Laboratories in 1960. The laser action took pace between two excited levels of neon, the function of the helium being to excite neon atoms to a high level—this process will be explained in detail in chapter three. A radiofrequency field of frequency 27 MHz was used as the source of excitation. The helium–neon laser is essentialy a four-level system (see fig. 1.17) which is much more efficient than the three-level type.

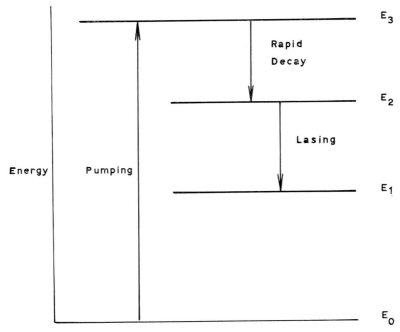

Fig. 1.17. The four-level laser.

151

Atoms in the ground state $E_0$ are excited to the highest level $E_3$ from which they descend, non-radiatively, to a metastable state $E_2$. Providing level $E_1$ is sufficiently high above the ground state, then it will be effectively empty and so a comparatively small population in $E_2$ is needed to ensure a population inversion between $E_2$ and $E_1$. Laser action can then take place between these levels. Clearly, therefore, four-level operation is much more efficient than three-level operation and hence continuous operation is much easier to obtain.

Javan's laser consisted of a quartz tube about 80 cm long having an inside diameter of 1·5 cm and filled with a mixture of helium at 1 torr pressure and neon at 0·1 torr pressure. The ends of the tube were terminated with a pair of flat parallel mirrors each being 98·9% reflecting. Oscillation was obtained at five wavelengths in the infra-red the strongest line being at 1·1523 $\mu$m with a power of 15 mW when 50 W of r.f. power were applied to the external electrodes.

By 1962 many research teams had started projects on lasers with the result that subsequently scores of different laser materials have been found with literally thousands of lasing lines available. Nevertheless of all these only a handful are to be found in common use as sources of light for various applications, but, before giving a description of these in chapter three, it will be helpful to obtain some quantitative ideas concerning laser action and to discuss the nature and properties of the resonating cavities involved.

# CHAPTER 3.2

## some laser theory

IN the last chapter the principles underlying laser operation were outlined in a general fashion. We shall now discuss some important quantitative relations which indicate the relative importance of the various parameters involved.

### 1. *Black-body radiation*

Except at a temperature of absolute zero, the atoms of a solid body are in motion; as a result collisions occur and some excitation takes place. On falling back to lower energy states a continuous spectrum of radiation is emitted—the hotter the body the greater the total quantity of radiation and the longer its average wavelength. In 1879, as a result of his measurements on radiation, Stefan proposed a law that related the total energy radiated from a body to its temperature. Stefan's law can be expressed by means of the following equation:

$$R = \sigma e T^4, \tag{2.1}$$

where $R$ is the total thermal radiation per unit surface area of the body, $\sigma$ is a constant called Stefan's constant, $T$ is the absolute temperature and $e$ is a number between 0 and 1 which depends on the nature of the emitting surface—this number is known as the emissivity.

By allowing a beam of radiation to fall onto various bodies and studying the radiation absorbed, Kirchhof, in 1895, showed that the ratio of the energy incident on the body to that absorbed by the body, $a$, was related to the emissivity quite simply by the equation:

$$e = a, \tag{2.2}$$

where $a$ is called the absorbtivity.

The radiation from the sun forms a continuous spectrum with dark lines superimposed on it; as explained in the last chapter, these occur because of absorption by the cooler outer layers of the sun. It is of considerable theoretical interest to study radiation from a body which has not undergone any absorption at any frequencies. In other words the body is a perfect emitter of radiation ($e = 1$) at all frequencies. Such an ideal body is called a black-body, and the radiation it emits is known as black-body radiation. In the laboratory this can be simulated by forming a hollow cavity within a solid body and drilling a small hole to join the cavity to the exterior. If the body is kept at a constant temperature then the radiation emitted from the hole will be a very good approximation to black-body radiation. By supplying energy to the body to make up for that lost from the cavity, a state of thermal equilibrium is established. The total radiation emitted by a black-body is from Stefan's law:

$$R = \sigma T^4. \tag{2.3}$$

It is of much more interest, however, to find out what amount of radiation can be expected at any particular frequency from a black-body. This can be done experimentally and graphs of the form shown in fig. 2.1 are obtained, these being plots of energy against frequency for various different temperatures.

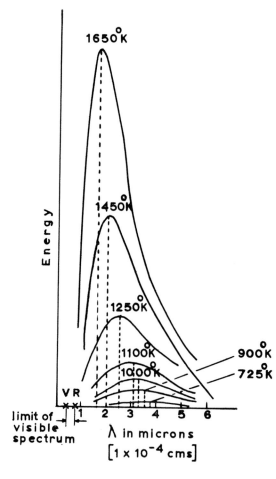

Fig. 2.1. Black-body radiation.

154

The area under each graph is proportional to the total energy emitted by the black-body at that particular temperature. As the temperature increases, the total energy increases in agreement with Stefan's law and the character of the radiation becomes shorter in wavelength. When a solid body, such as a piece of iron, is heated, at first it does not change in colour because nearly all the radiation emitted is in the far infra-red region of the spectrum to which the eye is not sensitive. As the temperature rises, the body becomes red hot and ultimately white hot. Under certain conditions it is possible for a body to be so hot that it appears blue; an example of this is a very hot type of star which has a distinctly blue appearance.

For some years before Planck's theory, many physicists, including Rayleigh and Jeans, had tried to work out a mathematical expression for the shapes of the curves shown in fig. 2.1 but without success. The advent of the quantum theory at last enabled a successful equation to be established. Without following the analysis in detail we shall indicate briefly the steps involved. First, imagine a cavity containing black-body radiation in thermal equilibrium. The radiation will take the form of standing waves within the cavity; a particular wavelength, or mode of oscillation, is possible if its nodes, or points where the amplitude of the wave is always zero, coincide with the walls of the cavity. Three different modes of oscillation are indicated in fig. 2.2.

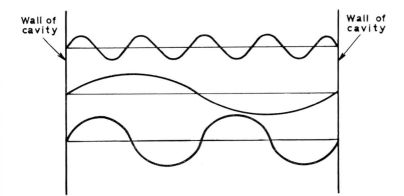

Fig. 2.2.  Black-body radiation modes.

From fig. 2.2 it can be deduced that a condition for a mode of oscillation to exist is:

$$\tfrac{1}{2}n\lambda = L, \tag{2.4}$$

where $\lambda$ is the wavelength, $n$ is a positive integer and $L$ is the length of the cavity.

The energy per frequency interval for a cavity of unit volume, which is what we wish to find, is obtained by multiplying the number of modes of oscillation per frequency interval by the average energy of a mode. In order to calculate the latter we must return to the Boltzmann distribution introduced in the last chapter, which tells us the relative population of different energy levels in a system in thermal equilibrium.

The population $N_1$ of an energy level $E_1$ is given by:

$$N_1 = A \exp\left(-E_1/kT\right), \tag{2.5}$$

where $A$ is a constant. The population $N_2$ of $E_2$ is similarly given by:

$$N_2 = A \exp\left(-E_2/kT\right). \tag{2.6}$$

If, therefore, two energy levels $E_1$ and $E_2$ exist where $E_2$ is greater than $E_1$, then

$$\frac{N_1}{N_2} = \frac{A \exp\left(-E_1/kT\right)}{A \exp\left(-E_2/kT\right)}, \tag{2.7}$$

from which it follows that

$$\frac{N_1}{N_2} = \exp\left(-\frac{E_2 - E_1}{kT}\right). \tag{2.8}$$

In some instances an atomic energy level really consists of a number of distinct energy levels which happen to coincide in their energy values. The energy level is then said to be degenerate; the symbol $g$ indicates the number of energy states superimposed on one level and the level is said to have a degeneracy of $g$. Therefore equations (2.5) and (2.6) are written more correctly as:

$$\frac{N_1}{g_1} = A \exp\left(-E_1/kT\right) \tag{2.9}$$

and

$$\frac{N_2}{g_2} = A \exp\left(-E_2/kT\right), \tag{2.10}$$

and therefore it follows that equation (2.7) is again written more correctly as:

$$\frac{N_1}{N_2} = \frac{g_1}{g_2} \exp\left(-\frac{E_2 - E_1}{kT}\right). \tag{2.11}$$

By using Planck's theory it is possible to calculate the average energy of each mode. According to Planck, the energy of a radiation field of frequency $\nu$ can take only the following values: zero, $h\nu$, $2h\nu$, $3h\nu$, etc., and from the Boltzmann distribution the number of oscillators having these energies is $A$, $A \exp\left(-h\nu/kT\right)$, $A \exp\left(-2h\nu/kT\right)$, etc. The average energy of a black-body radiation mode, $E_1$, is hence simply obtained by dividing the total energy by the total number of oscillators, i.e.

$$\epsilon = \frac{0A + h\nu A \exp\left(-h\nu/kT\right) + 2h\nu A \exp\left(-2h\nu/kT\right) + \cdots}{A + A \exp\left(-h\nu/kT\right) + A \exp\left(-2h\nu/kT\right) + \cdots}. \quad (2.12)$$

On simplifying:

$$\epsilon = \frac{h\nu \exp\left(-h\nu/kT\right)[1 + 2 \exp\left(-h\nu/kT\right) + 3 \exp\left(-2h\nu/kT\right) + \cdots]}{1 + \exp\left(-h\nu/kT\right) + \exp\left(-2h\nu/kT\right) + \cdots}. \quad (2.13)$$

The numerator is easily reduced by the binomial theorem and the denominator is a geometric progression and so equation (2.13) can be expressed in the much simplified form:

$$\epsilon = \frac{h\nu}{\exp\left(h\nu/kT\right) - 1}. \quad (2.14)$$

The number of modes per unit volume of the cavity per unit frequency interval is obtained by a straightforward procedure which can be found in physics textbooks, and is given by:

$$\frac{8\pi\nu^2}{c^3}, \quad (2.15)$$

and so the energy density or energy per unit volume $u_\nu$ of the cavity, per unit frequency interval, is given by:

$$u_\nu = \frac{8\pi h\nu^3}{c^3} \frac{1}{\exp\left(h\nu/kT\right) - 1}. \quad (2.16)$$

If $u_\nu$ is plotted, as a graph, against frequency $\nu$, then curves identical to those shown in fig. 2.1 are obtained, a different curve being obtained for each temperature.

An interesting comparison can be made between the radiation emitted from a black-body and that from a laser. A ruby laser gives an output of 6943 Å in the red part of the spectrum and is capable of giving pulses of about a megawatt per square centimetre with a line-width of about 0·1 Å. A black-body whose radiation peaks at 6943 Å would have to be at a temperature of 4174°K. In a region 0·1 Å wide about this peak the radiation flux is only about 16 mW cm$^{-2}$, which indicates the vast power a laser can give when compared to a black-body. To give a fairer comparison, however, if a ruby laser is made to work continuously then an output of 4 mW can be obtained from an initial electrical input of 850 W—the efficiency therefore being about 0·0005%. The total energy of the black-body radiation example given above is 1700 W cm$^{-2}$, so the efficiency is approximately 0·001%. The ruby laser, however, is a particularly inefficient one and most gas lasers, for example, are 0·01% efficient while the carbon dioxide laser can be as much as 20% efficient. Semiconductor lasers can reach efficiencies of 10%.

## 2. *The Einstein coefficients*

The lifetime of an atom is the average time it exists in an excited state before it makes a spontaneous transition to a lower energy state, which can also be thought of as the time taken to emit a photon.   As was stated in chapter one, an excited atom in free space has a lifetime of about $10^{-8}$ s.   Another way of visualizing this is to consider the reciprocal of the lifetime which is simply equivalent to the average number of spontaneous transitions from an excited state within a period of a second, and so in the case of the free state atom, $10^8$ atoms will make such transitions per second on average.   This transition rate is synonymous with transition probability—the larger the transition rate the greater the probability of transition.   The probability of a spontaneous emission occurring denoted by $A$, is called the Einstein $A$ coefficient.   It follows that

$$A_{21} = \frac{1}{t_{21}}, \tag{2.17}$$

the suffix 21 indicating a transition from an upper level labelled 2 to a lower level labelled 1.   $t_{21}$ is the lifetime and is synonymous with the $\Delta t$ introduced in Chapter 1.

In a similar fashion two Einstein coefficients can be defined for the cases of stimulated emission and absorption.   They are termed $B_{21}$ and $B_{12}$, again the order of the numbers indicates the initial and final states. The same letter $B$ is used for both $B$ coefficients as it is found that $B_{21}$ and $B_{12}$ are in fact identical.   By using the black-body radiation formula equation (2.16) we shall now give a proof of this and in addition derive a relation between the Einstein $A$ and $B$ coefficients.

In the two-level energy system shown in fig. 2.3 $N_2$ and $N_1$ are the populations of levels $E_2$ and $E_1$ respectively.   The total number of spontaneous downward transitions from level 2 to level 1 each second is:

$$N_2 A_{21}. \tag{2.18}$$

Photons of energy $h\nu$ arriving to interact with the system will either be absorbed or will stimulate emission.   In each case the greater the rate of arrival, which is proportional to the radiation density denoted by $\rho(\nu)$, the more likelihood of such transitions taking place.   For absorption the total number of absorptions taking place per second will be given by:

$$N_1 B_{12} \rho(\nu), \tag{2.19}$$

and for stimulated emission the total number of stimulated transitions from the upper level to the lower level will be:

$$N_2 B_{21} \rho(\nu). \tag{2.20}$$

Suppose now that the system is in thermal equilibrium.   This means that the total energy of the system must remain constant or, in other words, the number of photons absorbed per second must be equal to the

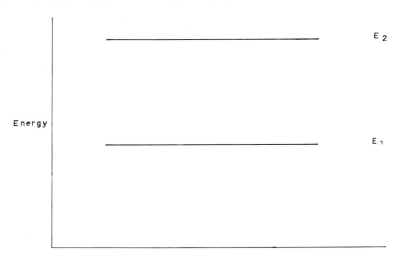

Fig. 2.3.  Two-level energy system.

total number emitted from the system by stimulated and spontaneous emission.  It thus follows from equations (2.18), (2.19) and (2.20) that

$$N_1 B_{12} \rho(\nu) = N_2 A_{21} + N_2 B_{21} \rho(\nu), \qquad (2.21)$$

from which

$$\rho(\nu) = \frac{N_2 A_{21}}{N_1 B_{12} - N_2 B_{21}}; \qquad (2.22)$$

however, by the Boltzmann distribution which applies to systems in thermal equilibrium we have:

$$\frac{N_2}{N_1} = \exp(-h\nu/kT), \qquad (2.23)$$

assuming that the degeneracy of each level is unity.

Substitution of equation (2.23) in (2.22) gives:

$$\rho(\nu) = \frac{A_{21}}{\exp(h\nu/kT)B_{12} - B_{21}}. \qquad (2.24)$$

Now an ideal two-level system such as we have been considering must result in a radiation density $\rho(\nu)$ which is identical to black-body radiation density.  By comparing the above equation (2.24) with the black-body radiation equation (2.16) it follows, because the two expressions for $\rho(\nu)$ must be identical, that

$$B_{12} = B_{21} = B \qquad (2.25)$$

and

$$\frac{A_{21}}{B} = \frac{8\pi h\nu^3}{c^3}. \qquad (2.26)$$

It was, in fact, the necessity of establishing this identity that led Einstein to introduce the concept of stimulated emission in 1917.

Equation (2.25) shows that the probability of stimulation down is equal to the probability of absorption which seems intuitively sensible if absorption is thought of as ' stimulation up '. Equation (2.26) has some important implications; the ratio $R$ of the rate of spontaneous emission to the rate of stimulated emission under conditions of thermal equilibrium is given by:

$$R = \frac{A_{21}}{\rho(\nu)B},\qquad (2.27)$$

which on substituting equation (2.24) into equation (2.27) becomes:

$$R = \exp\left(h\nu/kT\right) - 1. \qquad (2.28)$$

If $\nu$ is taken to correspond to the frequency of green light i.e. $0.5 \times 10^{14}$ Hz, then $R$ is found to be equal to about $e^{82}$ or roughly $10^{35}$! In other words, the likelihood of a stimulated emission taking place is completely negligible compared to the probability of a spontaneous emission. If a frequency corresponding to a microwave transition is taken, e.g. $10^9$ Hz, then $R$ becomes about $0.001$—a complete reverse in the probability situation. Radio waves and microwaves arise almost entirely from stimulated emission and so are always coherent. In all cases spontaneous emission manifests itself as undesirable noise within the system. To complete these numerical ideas it is found that the rates of spontaneous and stimulated emission become equal, i.e. $R = 1$ for a wavelength of about 60 $\mu$m in the far infra-red region of the spectrum.

In this section we have assumed that thermal equilibrium exists and under such conditions the possibility of stimulated emission in the visible region of the spectrum is entirely negligible. By creating a population inversion, the thermal equilibrium is effectively destroyed and considerable stimulated emission of visible light becomes possible.

## 3. *The threshold condition*

From the point of view of efficiency, it is clearly of importance to choose laser materials which require the minimum amount of energy before starting to lase. In practice this means finding out what population inversion must be achieved before all the losses in the system are overcome. A laser consists of an amplifying medium, usually a gas or a solid, placed between two mirrors which form an optical resonator. The losses in this system are replenished by induced transitions occurring within the amplifying medium. The total loss is due to a number of different processes, the most important of which are:

(1) Transmission, absorption and scattering by the mirrors.
(2) Absorption within the amplifying medium due to other energy levels—no system is an ideal two-level one.

(3) Scattering by optical inhomogeneities within the amplifying medium—a particularly important source of loss in the case of solid-state lasers where it is impossible to provide a perfect crystal.

(4) Diffraction losses by the mirrors—this important aspect is discussed further later.

All these losses can be included in one parameter which can be expressed as the lifetime of a photon existing within the laser cavity. The amplifying medium is assumed to be neutral, i.e. the photon will not undergo any processes which are concerned with the two-energy levels between which laser action takes place. Such a lifetime is denoted by the symbol $t_{\text{photon}}$. The reciprocal of this will be the total rate of loss of protons from the laser per second as a result of the four processes outlined above.

Consider a two-level energy system denoting stimulated transition rate for the time being by:

$$W' = \rho(\nu)B. \tag{2.29}$$

Substituting from equation (2.26) it follows that

$$\frac{W'}{A} = \rho(\nu)\frac{c^3}{8\pi h\nu^3}, \tag{2.30}$$

where the suffix 21 has been dropped from the term $A_{21}$.

It was mentioned previously that the laser line-width is very narrow indeed—much narrower, in fact, than the line-width of the atomic transition. The shape of the absorption curve is denoted by $g(\nu)$ and consideration will show that $g(\nu)$ can also be described as an emission curve or a frequency probability curve. We can define $g(\nu)d\nu$ as the probability that a given transition between energy levels will result in an emission or an absorption, of a photon whose energy lies between $h\nu$ and $h(\nu + d\nu)$. The curve $g(\nu)$ is normalized so that the total area under it is always unity. The effect of all this is that a photon of frequency $\nu$ will not, with absolute certainty, stimulate another photon of frequency $\nu$; it can only be stated that there will be a finite probability $g(\nu)d\nu$ that the stimulated photon will have a frequency which is between $\nu$ and $\nu + d\nu$. Similar arguments also apply to spontaneous emission and it is therefore necessary to replace $W'$ and $A$ in equation (2.30) by $W'g(\nu)d\nu$ and $Ag(\nu)d\nu$ respectively. $W'g(\nu)d\nu$ is now the rate at whch transitions take place, resulting in photons of frequency lying between $\nu$ and $\nu + d\nu$ due to a radiation density $\rho(\nu)$. $Ag(\nu)d\nu$ is the rate of spontaneous emission into the frequency interval lying between $\nu$ and $\nu + d\nu$. Equation (2.30) may be therefore rewritten to give:

$$W'g(\nu)d\nu = \frac{\rho(\nu)c^3}{8\pi h\nu^3}Ag(\nu)d\nu. \tag{2.31}$$

The radiation density $\rho(\nu)$ is simply related to the radiation flux or intensity $I(\nu)$ by:

$$I(\nu) = c\rho(\nu), \tag{2.32}$$

where $c$ is strictly speaking the velocity of light in the laser medium, and using the fact that

$$A = \frac{1}{t_{\text{spont.}}}, \tag{2.33}$$

where $t_{\text{spont.}}$ is the spontaneous emission lifetime. On integration of each side to give the total transition rate, $W(\nu)$, due to a monochromatic beam of radiation of frequency denoted by $I_\nu$, equation (2.31) becomes:

$$W(\nu) = \frac{c^2}{8\pi h \nu^3 t_{\text{spont.}}} g(\nu) I_\nu. \tag{2.34}$$

We are now in a position to write down the increase in intensity due to stimulated emission, but before doing so we must allow for any degeneracy in energy levels. Degeneracies of unity were assumed throughout the analysis of the Einstein coefficients. Suppose a degeneracy of $g_1$ is associated with the lower level $E_1$ and a degeneracy of $g_2$ with the upper level $E_2$ (this is standard notation and should not be confused with $g(\nu)$, the line-shape function). Bearing these degeneracies in mind, the net rate of change in energy due to stimulated transitions (both up and down) is obtained by multiplying the transition rate by the net population inversion and the photon energy $h\nu$. I.e.

$$h\nu \left( N_2 - N_1 \frac{g_2}{g_1} \right) W, \tag{2.35}$$

therefore the rate of increase in intensity may be expressed as:

$$\left( \frac{dI}{dt} \right)_{\text{gain}} = h\nu \left( N_2 - N_1 \frac{g_2}{g_1} \right) Wc. \tag{2.36}$$

A gain or increase in intensity is assumed because a population inversion is also assumed i.e.:

$$N_2 \geqslant N_1 \frac{g_2}{g_1}. \tag{2.37}$$

If this condition does not hold, the laser medium will simply absorb, which is what normally happens in any material without population inversion. The total loss rate is defined by means of the parameter $t_{\text{photon}}$ discussed above:

$$\left( \frac{dI}{dt} \right)_{\text{loss}} = \frac{I}{t_{\text{photon}}}. \tag{2.38}$$

162

Combining equations (2.36) and (2.38) we obtain a condition which clearly must be satisfied for laser action to take place:

$$\left(\frac{dI}{dt}\right)_{\text{gain}} - \left(\frac{dI}{dt}\right)_{\text{loss}} \geqslant 0 \qquad (2.39)$$

or

$$\left(N_2 - N_1 \frac{g_2}{g_1}\right) \geqslant \frac{8\pi t_{\text{spont.}} \nu^2}{c^3 g(\nu_0) t_{\text{photon}}}. \qquad (2.40)$$

This is the threshold population inversion required for oscillation near the line $g(\nu_0)$. For minimum threshold inversion the largest value of $g(\nu)$ is taken which occurs, of course, at the line centre $\nu_0$.

A number of important indications concerning the choice of suitable laser materials can be drawn from this important equation, but first it is necessary to express $g(\nu_0)$ more explicitly.

The shape of the curve $g(\nu)$ will depend on the physical processes which give rise to it. These processes can be divided into two groups, which give rise to what is termed inhomogeneous and homogeneous broadening.

Inhomogeneous broadening happens because a particular frequency on the $g(\nu)$ curve can be associated directly with a particular atom in the laser medium. If the medium is a gas, then the broadening is caused by motion of the atoms—each atom ' sees ' a radiation field of different frequency, depending on the velocity of its motion. For a solid state material inhomogeneous broadening is caused by the close proximity of atoms whose strong electric fields cause the energy levels associated with each atom to be disturbed and thus giving a spread in line-width.

Homogeneous broadening, on the other hand, is a result of probability. For instance, pressure broadening results in a line of finite width because the presence of other atoms results in collisions which have the effect of altering the lengths of the wave-trains undergoing transitions. The collisions are a random process so it is not possible to associate a given frequency with a given atom—only a probability can be specified. Thus, unlike the case of inhomogeneous broadening, removal of a particular atom in the system will cause a lowering on the overall height of the $g(\nu)$ curve rather than leave a gap at a particular frequency, which is what happens with an inhomogeneously broadened line. As, by definition, the area under the $g(\nu)$ curve is always normalized to unity the line-width will narrow slightly. As more and more atoms are removed the line-width will get smaller and smaller, as is to be expected, because the probability of one atom colliding with another becomes less.

Inhomogeneous and homogeneous broadening results in lines, i.e. $g(\nu)$ curves, of different shapes. An inhomogeneously broadened line is Gaussian in shape and a homogeneously broadened line has what is known as a Lorentzian shape. The equations of these curves are given below and their relative shapes indicated in fig. 2.4.

For a Gaussian curve:

$$g(\nu) = \frac{2(\pi \ln 2)^{1/2}}{\pi \Delta \nu} \exp \left[ -\left( \frac{\nu - \nu_0}{\Delta \nu / 2} \right)^2 \right]; \tag{2.41}$$

$$\therefore \quad g(\nu_0) = \frac{2(\pi \ln 2)^{1/2}}{\pi \Delta \nu} \tag{2.42}$$

$$\Delta \nu = \frac{2\nu_0}{c} \sqrt{\left( \frac{2kT \ln 2}{m} \right)} = 7 \cdot 162 \times 10^{-7} \times \nu_0 \sqrt{\frac{T}{m}}, \tag{2.43}$$

where $m$ is the mass of the atom.

For a Lorentzian curve:

$$g(\nu) = \frac{\Delta \nu}{2\pi[(\nu - \nu_0)^2 + (\Delta \nu / 2)^2]}; \tag{2.44}$$

$$\therefore \quad g(\nu_0) = \frac{2}{\pi \Delta \nu}, \tag{2.45}$$

$$\Delta \nu = \frac{1}{\pi \tau}, \tag{2.46}$$

where $\tau$ is the time interval between each collision.

Substituting the value of $g(\nu_0)$ into the threshold equation (2.40) for the Lorentzian line (it is not necessary to consider the Gaussian value as from equations (2.42) and (2.45) it can be seen to be a constant multiple of the Lorentzian) we have:

$$N_2 - N_1 \frac{g_2}{g_1} \geqslant \frac{4\pi^2 \nu^2 \Delta \nu}{c^3} \left( \frac{t_{\text{spont.}}}{t_{\text{photon}}} \right) = \Delta N_{\text{c}}. \tag{2.47}$$

For the smallest possible population inversion necessary for gains to exceed losses it can be seen from equation (2.47) that the following requirements for suitable laser media are:

(1) A material must be selected in which the lower level population $N_1$ can be kept small. A fast relaxation rate out of the level to other lower levels is necessary.

(2) The atomic line-width $\Delta \nu$ should be as small as possible. Cooling of the laser material will clearly be of advantage here.

(3) The spontaneous emission $t_{\text{spont.}}$ out of the upper level should be kept as short as possible but that out of the lower level to other levels should also be short, in fact shorter than $t_{\text{spont.}}$, otherwise the lower level will become 'clogged up' and population inversion will be impossible.

(4) $t_{\text{photon}}$ should be as long as possible; this means that the output mirror should be of high reflectivity and losses in medium and the cavity (due to such things as optically inhomogeneous materials, poor quality mirrors etc.) should be as small as possible. In practice the choice of

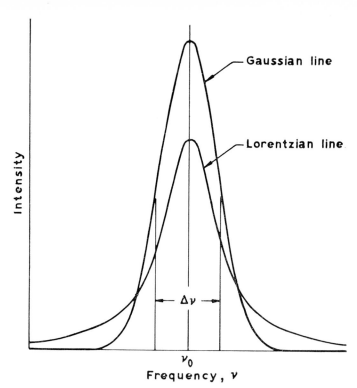

Fig. 2.4. Gaussian and Lorentzian line-shapes.

reflectivity of the output mirror is a compromise between gain within the cavity and output power. Low gain laser lines require high reflectivity output mirrors (typically 1–3% transmission) while high gain lines give most output power with lower reflectivity mirrors (typically 5–10% transmission).

(5) In the case of the Lorentzian line $\Delta\nu$ is independent of frequency as indicated in equation (2.46) but inversely dependent on $\tau$, the time between collisions. This implies that high gas pressures should be avoided otherwise $\Delta\nu$ will be excessively large. The minimum population inversion is proportional to $\nu^2$.

In the Gaussian (inhomogeneous) case, the line-width is proportional to the frequency, cf. equation (2.43), so that on substitution for $g(\nu_0)$ from equation (2.42) in equation (2.40) it can be seen that the minimum population inversion is proportional to $\nu^3$ which implies that the prospects for the construction of X-ray or $\gamma$-ray lasers are poor.

### 4. *Minimum pumping power*

It is instructive to get some quantitative idea of how much population inversion must be obtained for three and four-level lasers respectively. Energy levels representing these two types are shown in fig. 2.5.

165

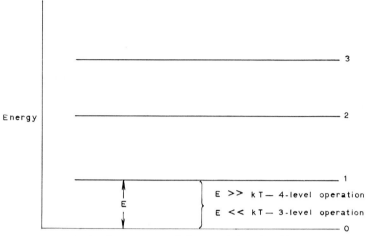

Fig. 2.5. Comparison of three and four-level lasers.

(a) *Four-level case*

Here $E \gg kT$ and hence levels 1, 2 and 3 start off effectively empty, hence only a few atoms need to be pumped into level 2 to achieve an inversion with respect to level 1. For the threshold of oscillation $N_2$ must be equal to $\Delta N_c$ the critical population inversion.

It follows therefore that

$$P_{\min} = \frac{\Delta N_c h\nu}{t_{\text{spont.}}}. \tag{2.48}$$

Substituting for $N_c$ from equation (2.47) we have:

$$P_{\min} = \left[ \frac{4\pi^2 \nu^2 \Delta\nu}{c^3} \left( \frac{t_{\text{spont.}}}{t_{\text{photon}}} \right) \right] \frac{h\nu}{t_{\text{spont.}}} \tag{2.49}$$

$$= \frac{4\pi^2 h\nu^3 \Delta\nu}{c^3 t_{\text{photon}}}. \tag{2.50}$$

(b) *Three-level case.*

Referring to fig. 2.5, in this case, $E \ll kT$ and the population of level 1 is about half the total number of atoms present; the other half being in level 0. Therefore to satisfy the threshold condition it is necessary that $N_2 \sim N_1 + \Delta N_c$ and so the required population inversion for a three-level laser must be much greater than that for a four-level laser by a factor $\Delta N_c + N_1/\Delta N_c$ or $1 + N_1/\Delta N_c$. Normally $N_1$ is very much greater than $\Delta N_c$ so the factor becomes $N_1/\Delta N_c$ or $N/2\Delta N_c$, where $N$ denotes the total number of atoms present, as $N_1$ is roughly half the total number of atoms present.

As a natural example of the difference between a three-level and four-level laser, consider the case of a typical solid-state laser where for a cubic centimetre of laser material, $\Delta N_c$ is $10^{16}$ and $\frac{1}{2}N$ is about $10^{18}$. For a four-level system the threshold condition is $N_2 = \Delta N_c$ or 5% of the total number of atoms. For a three-level system, however, the critical value of $N_2$ will be $10^{18} + 10^{16} \sim 10^{16}$, i.e. for a four-level laser approximately $10^{16}$ atoms will have to be pumped into level 2 and for a three-level laser the number required will be approximately $10^{18}$, indicating that the pumping power will have to be about a hundred times greater for a three-level laser.

## 5. *Resonators*

Gain in a laser is enhanced by placing two mirrors facing each other with the active laser medium in between. This arrangement of mirrors is essentially a Fabry–Perot interferometer. Once a stimulated transition takes place inside the laser medium, and a photon is emitted in a direction parallel to the axis of the laser, then the amplitude of the wave is increased by further stimulated emissions—the wave being returned, by means of the mirrors, back into the active medium and so increasing still further in intensity. A laser without any mirrors at each end is an amplifier; with mirrors it becomes an oscillator. Any oscillator is essentially a device which returns some of the output from some sort of system back to the input, such that the return energy is in phase with the input energy. Oscillators can be mechanical, electrical or optical and in the case of the laser, which is an optical oscillator, the feedback is achieved by means of the laser mirrors. In order that some energy can be obtained as an output one mirror is made slightly transmitting and is therefore called the output mirror. The other mirror should, ideally, be 100% reflecting, but in practice it is impossible to make a mirror as good as this. Reflectivities of over 99% are obtained by evaporating alternate layers of materials, whose dielectric constants are different, on to a silica substrate (commonly used materials are zinc sulphide and cryolite which are easily deposited and produce soft coatings—if hard coatings are required the titanium dioxide and silicon monoxide are often used). The evaporation process begins and ends with the deposition of a material having the higher dielectric constant (with the consequence that an odd number of layers is always evaporated). For the highest reflectivity twenty-one to twenty-five layers are used. Output mirrors of lower reflectivity, and hence higher transmission, are obtained by depositing fewer layers, e.g. an approximately 5% transmitting mirror has seven layers, and an approximately 10% transmission requires only five. Throughout the analysis which follows it is assumed that the laser mirror, which is not the output mirror, is a perfect reflector.

In Javan's first helium–neon laser the mirrors were placed inside the tube containing the mixture of helium and neon. Ruby lasers have mirrors which can be made by simply polishing the ends of the ruby rod flat and parallel and evaporating a layer of silver onto them. Positioning the mirrors inside the tube of a gas laser is inconvenient for the following reasons: (*a*) adjustment of the mirrors in order to establish correct alignment is very difficult; (*b*) gas purity may be an important consideration and tubes are often baked before filling in order to drive off foreign gases—a process to which dielectric mirrors are not particularly amenable. The only advantage of internal mirrors is that losses do not occur as a result of transmission through the end of the tube. However, such losses can be greatly reduced by terminating the tube containing the gas or cutting the ruby rod so that a particular angle is made with the optic axis of the laser called the Brewster angle. This results in reflection losses for one direction of polarization being very low indeed. The two pieces of glass which form the terminations to laser tubes are called Brewster windows and the use of these naturally results in the output of the laser being highly polarized. A typical gas laser tube with Brewster windows and mirrors in position is shown in fig. 2.6.

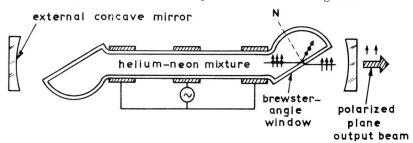

Fig. 2.6. The components of a gas laser (a helium–neon laser). (After Yariv, *Proc. I.E.E.E.*, 1963.)

We shall now derive an expression for $t_{photon}$ in terms of the reflectivity of the output mirror. It will be assumed that not only is one laser mirror 100% reflecting, i.e. it has a reflectivity $R$ which is equal to 1, but also that the light transmitted through the output mirror is the only source of loss from the cavity, i.e. no scattering, absorption etc. This is, of course, not so in practice but will give a good approximation to the magnitude of $t_{photon}$. It will be remembered that the reciprocal of $t_{photon}$ is effectively the rate of loss of photons per second. A loss will occur, on average, after $1/1-R$ passes up and down the laser cavity. If, for example, the reflectivity of the output mirror $R = 1$, then the photon will not escape into the output until an infinite number of passes have occurred, i.e. it will never escape and, if $R = 0$, i.e. the output mirror is perfectly transmitting, then not more than one pass will occur on average before loss. If $d$ is the cavity length, the time taken to make $1/1-R$ passes is given by dividing the total distance travelled $2d/1-R$ by the velocity of light $c$. The reciprocal of this must be equal to the rate of loss per second which is equal to the reciprocal of $t_{photon}$ and so:

$$\frac{1}{t_{photon}} = \frac{c(1-R)}{2d}; \qquad (2.51)$$

$$\therefore \quad t_{photon} = \frac{2d}{c(1-R)}. \qquad (2.52)$$

For a typical ruby laser the cavity length $d$ may be 10 cm and the reflectivity of the output mirror $R$ may be 0·98 which, on substitution into equation (2.52), results in a value of $t_{\text{photon}}$ of about $3 \times 10^{-8}$ s. Therefore, assuming a neutral medium where no absorption or stimulated or spontaneous emission takes place, a photon will be lost, on average, $3 \times 10^{-8}$ s after coming into existence by means of an energy level transmission. The cavity length of 10 cm implies that fifty passes will occur, again on average, before the photon leaves the laser.

All resonating systems can be characterized by a $Q$ factor or quality factor $Q$ defined as the ratio of the frequency of oscillation to the line-width of the output. A high quality resonator which has a large $Q$ factor will therefore have an output whose line-width is very narrow. We have:

$$Q = \frac{\nu}{\Delta \nu}. \tag{2.53}$$

It can be shown that an equivalent expression for $Q$ is given by:

$$Q = \frac{2\pi\nu \text{ (energy stored in resonator)}}{\text{energy lost per second from the resonator}}. \tag{2.54}$$

A typical electrical oscillator will have a $Q$ of the order of 100—its line-width will thus be 0·01 of its frequency. Let us now consider the case of a Fabry–Perot interferometer with a photon between its mirrors and no laser medium present. The $Q$ is given by equation (2.54) to be:

$$Q = 2\pi\nu \frac{h\nu}{h\nu/t_{\text{photon}}}; \tag{2.55}$$

$$\therefore \quad Q = 2\pi\nu \, t_{\text{photon}}; \tag{2.56}$$

so by substitution from equation (2.52) for $t_{\text{photon}}$ we get:

$$Q = \frac{4\pi\nu d}{c(1-R)} \tag{2.57}$$

and therefore

$$Q = \frac{4\pi d}{\lambda(1-R)} \tag{2.58}$$

by using equation (1.4).

For the case of a Fabry–Perot interferometer whose mirrors are 100 cm apart, one mirror being 2% transmitting and the other 100% reflecting, i.e. a typical mirror system used for a helium–neon laser, then assuming no other losses, we have: $\lambda = 6328$ Å, $d = 100$ cm and $R = 0.98$. Substituting into equation (2.58) gives a $Q$ whose value is $10^9$. That is to say, the line-width is $10^{-9}$ of the frequency, i.e. about $10^{-9} \times 5 \times 10^{14}$ Hz or 0·5 MHz. It is thus apparent that an optical resonator is intrinsically of much higher quality than an electrical oscillator, or for that matter a microwave oscillator. However, this is not the end of the story, because, as mentioned in Chapter 1, in the case of a laser, part of the volume between the mirrors is filled with an amplifying medium with the result that the denominator of equation (2.54) is reduced to such an extent that the quality factor becomes even greater which, in turn, means that the line-width becomes even smaller. This is the reason why a laser is capable of giving outputs whose line-widths are only a few hertz. A laser is described as an active resonator while an ordinary Fabry–Perot interferometer, with no active medium between its mirrors, is described as a passive resonator. Although the latter can act as a very narrow line-width filter to a beam of incident light, the transmitted beam can never have as narrow a line-width as that emitted by a laser.

We now turn our attention to the actual frequencies, or modes of oscillation as they are called, which are obtained in the output beam from a laser. In a microwave oscillator such as a klystron, the resonant cavity is of similar size to the wavelength of the oscillations, but this is not the situation in the laser where the wavelength of light is minute in comparison with the size of the resonator. This has the effect of making a large number of modes (different frequencies) of oscillation possible. These are the so-called resonant modes of a Fabry–Perot interferometer. Figure 2.7 shows one such mode—a mode can only exist where an

Fig. 2.7. An axial mode.

integral number of half wavelengths is equal to the length of the cavity so that each end face of the cavity corresponds to a node, i.e.

$$n\frac{\lambda}{2} = d. \qquad (2.59)$$

Each of the above modes is called an axial mode because its direction of propagation is parallel to the axis of the resonator. The frequency separation $\Delta\nu$ between each mode will be constant and can be obtained by writing equation (2.59) for two adjacent modes as follows:

$$n\frac{\lambda_1}{2} = d, \qquad (2.60)$$

$$(n+1)\frac{\lambda_2}{2} = d. \qquad (2.61)$$

By eliminating $n$ and using equation (1.5), equations (2.30) and (2.61) reduce to:

$$\Delta\nu = \frac{c}{2d}; \qquad (2.62)$$

for a cavity 100 cm long $\Delta\nu$ is 150 MHz.

Only modes having sufficient gain to overcome all the losses can oscillate which means in practice that the laser output consists of several modes because of the finite width of the atomic transition line. A typical width for a Doppler broadened gas line is about 1000 MHz. Figure 2.8 illustrates this situation.

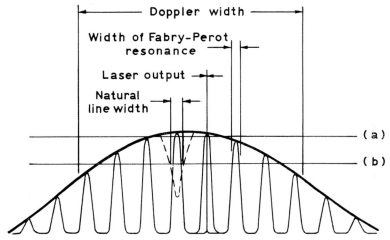

Fig. 2.8. Axial modes and line-widths.

Horizontal lines can be drawn to indicate the gain of the laser and the height of this line will ultimately govern how many axial modes actually oscillate. The higher the horizontal lines the greater the loss or the lower the gain. Loss line (a) would result in two lines oscillating and loss line (b) corresponding to a higher gain, would result in six lines oscillating. Reducing the gain will reduce the number of axial modes in the output, and also at the same time reduce the total power output which is proportional to the total area under each of the oscillating mode curves. Figure 2.8, besides indicating the Doppler width and the width of the Fabry–Perot resonator, also shows the natural line-width of the atomic transition and the line-width of the laser output. Unless special precautions are taken it is very difficult to obtain an output which consists of only one axial mode. This difficulty arises because if the gain is made sufficiently small, for instance, to allow only one axial mode to oscillate (the axial mode having the highest gain), then the intensity of the output will fluctuate. This is caused by shifts in the actual frequencies

171

of the axial modes resulting from changes in the length of the cavity which arise as a result of inevitable thermal and vibrational movements. Most lasers oscillate on several axial modes so that as the axial mode at the far end of the gain curve falls in intensity another at the other end of the gain curve starts to increase in intensity and so the total intensity of the output is kept fairly constant. It is the presence of more than one axial mode that reduces the coherence length of the laser which, it will be remembered from Chapter 1, is proportional to the overall line-width of the output. Reducing the number of axial modes gradually increases the coherence length; but when the gain is reduced so that the three axial modes give way to one (it is assumed that the two modes on each side of the central mode are equal in gain so three modes change to one as the gain is reduced) the line-width is reduced from 150 MHz to a few thousand cycles with a consequent enormous increase in the coherence length.

The output of a laser is often a beam which, when enlarged by projection onto a screen by means of a lens, is seen to exhibit a regular distribution of intensity across its width. Quite often enlargement is not even necessary for these patterns to be apparent. They arise from modes of oscillation which are the result of photons travelling up and down the laser cavity at small, but non-zero angles, with the optic axis. These modes are called transverse modes and a photograph of a typical pattern is shown in fig. 2.9.

Transverse modes require a higher gain to oscillate than axial modes. A laser which is not oscillating in any off-axis transverse modes but purely in axial modes gives a uniphase output. Transverse modes are often a nuisance because, by their nature, they result in an output beam which has a wider divergence than a uniphase beam. However, they can be eliminated by making the losses so high within the cavity that only axial modes have sufficient gain to oscillate and in practice this is done merely by placing apertures within the laser cavity. Each axial mode frequency has a large number of transverse modes of differing intensity patterns are associated with it. The frequencies of each set of transverse modes are, in general, very close to the frequency of the associated axial mode.

So far the actual shape of the reflecting surfaces of laser mirrors has not been discussed. Up to now they have been tacitly assumed flat, but plane mirrors are, in practice, very rarely used for reasons which will now be given.

In diagram 2.10 a a plane wave-front AA' is shown propagating in some direction; because of the wave nature of light such a beam will tend to spread out and its wavefront will become spherical—this process is called diffraction.

AB and A'B' are lines drawn normal to the wave-fronts. The uniphase wave obtained from a laser decreases in intensity across its width and A, A', B, B', C, and C', represent the points where the intensity

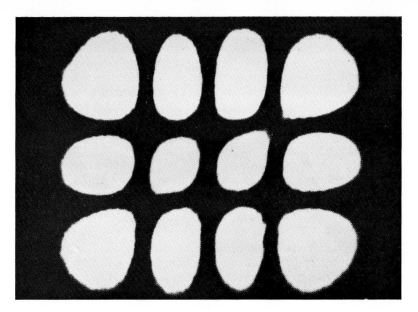

Fig. 2.9.   A typical transverse mode pattern.

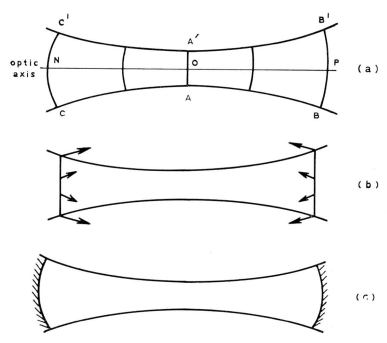

Fig. 2.10.   Curved mirrors.

of the beam has fallen to $1/e^2$ of its value on the axis. (Figures 2.10 *a*, 2.10 *b* and 2.10 *c* are cross-sections of the beam. In three dimensions the beam is obtained by rotation about the axis NOP.) It is obvious that if a plane mirror is placed at CC′ to form one mirror of the laser cavity, for example, then some of the reflected light will be diverged out of the system, (see fig. 2.10 *b*) and only part of it will reach a similar mirror placed at BB′. This diffraction loss is greatly reduced by making the mirror CC′ and BB′ curved so that they exactly match the curvature of the beam (see fig. 2.10 *c*). Some diffraction loss will always occur, of course, unless the mirrors are almost infinitely wide. Laser mirrors often have radii of curvatures of several metres. Such curved mirrors are to be preferred for the following reasons:

(1) Less light is lost by diffraction if curved mirrors are used rather than plane and so less gain is required for one equivalent output power.

(2) Curved mirrors are easier to align—they need to be aligned to an accuracy of only about one minute of arc whereas flat mirrors must be aligned to an accuracy of about one second of arc.

## 6. *Q-switching*

Pulsed lasers can be made to give extremely high power outputs by making them deliver their energy in a much shorter time than would otherwise be normal. The ruby laser, for example, normally delivers its energy over a time period of about $10^{-3}$ s, but by a process known as $Q$-switching this time can be reduced to almost $10^{-9}$ s (a nanosecond). The total energy delivered by the laser remains constant but the power, which is, of course, related to the energy in the following way:

$$\underset{\text{(watts)}}{\text{power}} = \frac{\text{energy delivered (joules)}}{\text{time (seconds)}}, \qquad (2.63)$$

increases by a factor of $10^6$. Thus a ruby laser, which has an energy output in one pulse of 10 joules, gives a power of a kilowatt for a millisecond. By $Q$-switching the energy can all be delivered in 10 ns and so the power output is $10^9$ W (one gigawatt). Under these conditions a small ruby laser can deliver as much power as the entire British National Grid system—but only for a few nanoseconds!

$Q$-switching, as its name implies, involves adjusting the $Q$ of the laser cavity so that feedback by the mirrors is suppressed and no depletion of the population of the upper energy level is permitted until its population is as high as possible. The laser is pumped with the resonator kept at a very low $Q$ value; the $Q$ of the cavity is then suddenly increaed, allowing gain from stimulated emission to take place, with the result that the laser energy escapes in a very short highly intense pulse. For obvious reasons, this process is also known as $Q$-spoiling or giant-pulse operation.

$Q$-switching is carried out by placing a shutter, which may take various forms, in the cavity and so effectively eliminating the resonator from the laser medium. The laser is pumped and the shutter very rapidly removed so restoring the cavity to the system. Two important requirements for $Q$-switching are: (*a*) the rate of pumping must be faster than the spontaneous decay rate from the upper energy level otherwise it will empty faster than it can be filled and population inversion will not be achieved; (*b*) the $Q$-switch must switch rapidly in comparison with the build-up in stimulated emission otherwise the latter will tend to be a gradual process and a longer pulse time than necessary will be obtained. In practice a $Q$-switch time of the order of 10 ns is preferable.

There are many ways of attaining $Q$-switching operation. The following methods work well and are to be found in common use:

(*a*) Rotating mirror method. A mirror or a prism is rapidly rotated within the cavity and only for the instant when the mirror is correctly aligned with respect to the optic axis will a high $Q$ exist. The pumping flash must obviously be synchronized with the mirror rotation. This type of switch is compact, reliable, cheap

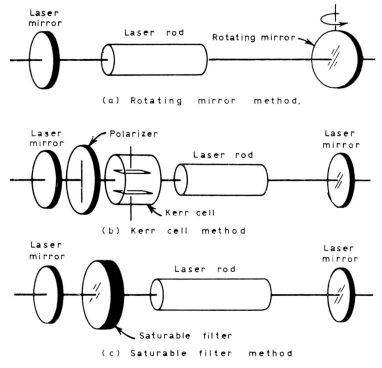

(a) Rotating mirror method.

(b) Kerr cell method

(c) Saturable filter method

Fig. 2.11. Three methods of $Q$-switching.

175

and simple and also has the advantage that it can be used in the infra-red. It does, however, tend to be slow in switching—a mirror rotating at 30 000 r.p.m., for example, switches in about 1000 ns.

(b) Kerr-cell method. This is an electro-optical $Q$-switch and depends on the fact that an electric field applied to a Kerr-cell (for example a container filled with nitrobenzine) changes the direction of polarization of a beam of light. As a $Q$-switch it is used to rotate the plane of polarization to be coincident with that of a polarizing filter placed in the cavity. When the electric field is switched on the cavity changes from one of high loss to one of low loss. This can be achieved in 10 ns but pumping synchronization is again necessary.

(c) Photochemical method. A photochemical substance, such as a saturable dye, forms a very convenient $Q$-switch; it has the great advantage that pumping synchronization is not required. A thin film of dye is placed in the cavity and when the intensity builds up to a sufficiently high level the dye is rapidly bleached, increasing the $Q$ of the cavity, and a giant pulse is emitted. Uranium doped glass can be used as the dye or a cell filled with cryptocyamine—both of these can be used over and over again without replacement as the bleaching action is only temporary.

Figure 2.11 shows the arrangement of components to produce the three types of switches described.

## principal laser types

THIS chapter will be devoted to descriptions of the most common type of laser, all of which have been used for a variety of applications.

### 1. *Solid-state lasers*

#### (a) *The ruby laser*

The first laser to be constructed used ruby as the active medium. A ruby laser, when it is $Q$-switched, still provides a most powerful and useful source of coherent light. The main components of the ruby laser are shown in fig. 1.16. The ruby crystal, is usually 1 mm to 2 cm in diameter and between 5 cm and 20 cm long and consists of alumina, (aluminium oxide $Al_2O_3$) to which has been added a small proportion (about $0.05\%$) of chromium which gives the crystal its characteristic pink or, if strongly doped, red colour. The chromium in the form of $Cr^{3+}$ ions provides the actual energy levels between which the stimulated emission essential to laser action takes place. The ruby crystal can be thought of as a ' gas ' of chromium ions. Figure 3.1 shows the energy level system for $Cr^{3+}$ in the host $Al_2O_3$.

The energy scale is given in centimetres, where $8066 \text{ cm}^{-1} = 1 \text{ eV}$. An explanation of the way in which the energy levels are labelled should be obtained from a text-book on spectroscopy; however, an understanding of the nomenclature is not necessary as far as this chapter is concerned.

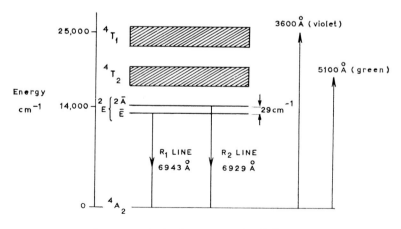

Fig. 3.1.   Ruby laser energy level diagram.

A percentage by weight of 0·05 means that there are about $1·62 \times 10^{19}$ chromium ions in each cubic centimetre of crystal. An xenon flash-tube similar to a photographic electronic flash-gun is used to excite or pump the ruby, the shape and position of which is most important if maximum efficiency is to be obtained. This follows because the basic three-level character of the ruby laser implies that at least half the ions in the $^4A_2$ ground level must be pumped up to the $^2E$ levels before laser action is possible. The flash-tube can take the form of a helical tube as in fig. 1.16, wound round the ruby rod with a coaxial cylindrical reflector. A better method is to arrange a straight flash-tube next to the ruby at the centre of the reflecting cylinder or to use an elliptical reflector with the ruby rod and flash-tube at the foci $f_1$ and $f_2$ as shown in fig. 3.2. Both of these arrangements are also easier to cool.

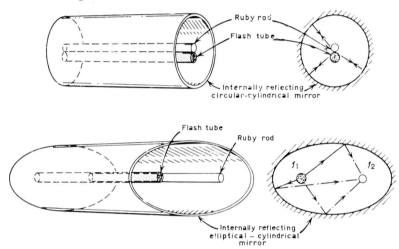

Fig. 3.2. Ruby laser flash-lamp reflectors. (After Harris, *Wireless Wld*, 1963.)

The absorption spectra for ruby peaks in the violet and green regions of the spectrum corresponding to the broad $^4T_1$ and $^4T_2$ levels shown in the energy level diagram for $Cr^{3+}$ fig. 3.1. These broad bands help to make the pumping by the ' white ' light from the flash-tube more efficient in comparison, for example, with pumping into a single line. The input energy to the flash-tube is provided by discharging a capacitor of 50 to 1000 $\mu F$ through a tube filled with the gas xenon to a pressure of 150 torr. The capacitor is charged up and the application of a high voltage pulse ionizes the gas and hence triggers its breakdown, producing a flash of pumping light of typically 500 to 1000 joules in energy and lasting for about 1 ms. The laser output has a power of about a kilowatt lasting for approximately a millisecond, consequently the output energy is of the order of a joule and so the efficiency is very much

178

less than $0.1\%$ since most of the energy supplied by the capacitor is dissipated as heat.

After pumping to the broad absorption levels, atoms drop down to the very closely spaced $^2E$ levels, which are in fact only 29 cm$^{-1}$ apart. The transition from the lower $\bar{E}$ levels to the ground state $^4A_2$ is called the $R_1$ transition and corresponds to a wavelength of 6943 Å. The transition from the upper $2\bar{A}$ level to the ground state, on the other hand, corresponds to a wavelength of 6929 Å and is termed the $R_2$ transition. Under normal conditions the threshold for the $R_1$ transition is lower and in addition the spontaneous lifetime of the $R_1$ line is greater than the thermal relaxation time for energy transfer between the $2\bar{A}$ and $\bar{E}$ levels with the result that the $R_1$ line dominates.

The output from the ruby laser commences approximately 0.5 ms after the start of the pumping flash and so lasts about half a millisecond (unless, of course, the laser is $Q$-switched). Once started, stimulated emission rapidly depopulates the upper lasing levels—much faster than the pumping rate can supply atoms and so the laser process has to pause and 'wait' until the population inversion is again achieved. The effect of this is to give an output which consists of a large number of spikes, each spike lasting about a microsecond as shown in fig. 3.3.

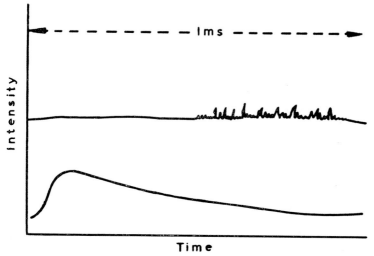

Fig. 3.3. Ruby laser and flash-lamp outputs versus time.

As a consequence of the large amount of heat dissipated by the flash-lamp, the laser becomes hot with the result that a limit must be set on the pulse repetition rate if overheating and damage are to be avoided. Only one pulse every few minutes is feasible if the laser is air cooled but by circulating water around the laser an increase in repetition rate to several pulses a minute is possible.

179

Ruby lasers lend themselves particularly well to $Q$-switching by which very high power outputs, even as high as 30 Gw, can be achieved for very short pulse times of the order of 10 ns.

The mirrors for ruby lasers can take the form of roof prisms either mounted externally or made by shaping the ends of the laser rod. Such roof prism resonators are extremely easy to align. Alternatively the flat ends of the ruby rod may be coated with layers of dielectric materials to produce very efficient mirrors. When it is desired to separate the mirrors from the ruby rod, which is necessary if $Q$-switching or discrimination against off-axis transverse modes by large mirror separation is required, then the ends of the rod are cut at the Brewster angle to minimize reflection losses.

### (b) The neodymium laser

The only other solid-state laser to compete with the ruby laser employs the energy levels of the rare earth element neodymium in the form of triply charged $Nd^{3+}$ ions. The host material can be a variety of substances of which calcium tungstate, glass or yttrium aluminium garnate (YAG) are commonly used. Glasses, owing to the considerable amount of development work which they have enjoyed, have the great advantage of high optical homogeneity and so offer more resistance to the damage which is always a problem with the high power densities associated with the solid-state lasers especially when they are $Q$-switched. They are also relatively cheap. An important difference between the ruby and neodymium laser is that the latter operates on four levels and so is inherently more efficient. The output is in the infra-red and consists of three principal wavelengths 0·914 $\mu$m, 1·06 $\mu$m and 1·35 $\mu$m. The 1·06 $\mu$m line is the most powerful and each of the wavelengths consists of a number of very closely spaced lines which arise through splitting of the neodynium lines by the internal electric field of the host material. For low-power pulsed and continuous operation calcium tungstate is used while for high-power pulsed operation neodymium in glass is preferable although it suffers from a very large spectral line-width of the output, typically 100 Å.

In the case of calcium tungstate the electric field effecting each neodymium ion is strong because a trivalent neodymium ion occupies a lattice position previously occupied by a divalent calcium ion. This effect is overcome by a process known as charge compensation in which another singly charged ion, such as sodium, is added to the material. A better solution is to use (YAG) as the host crystal; yttrium is trivalent and so no charge compensation is necessary. It also has better infra-red pumping bands and so the threshold of laser action for neodymium in YAG is about three times lower than that of Nd in calcium tungstate. Neodymium in YAG lasers can be operated continuously with power outputs of several watts.

An interesting method for obtaining high-power continuous outputs

180

in the green is to pass the 1·06 $\mu$m wavelength from a Nd–YAG laser into a crystal of either lithium niobate, KDP (potassium dihydrogen phosphate) or barium sodium niobate. These crystals have the effect of producing a harmonic at double the frequency of that of the input and in this way several hundred milliwatts at 5300 Å have been obtained continuously.

## 2. *Gas lasers*

Owing to a wide choice of wavelengths and outputs which are a close approach to an ideal coherent source of light, gas lasers are perhaps the most useful and certainly the most ubiquitous type of laser. Each of the three major types works on a slightly different basis. In the case of the helium–neon laser, a transition between unionized atomic states is involved, the argon laser utilizes a transition between ionized states and the carbon dioxide laser uses levels which arise from molecular rotation and vibration.

The two most significant mechanisms whereby an atom in a gas can be excited to a higher energy state are known as collisions of the first and second kind. Collisions of the first kind involve the interaction of an energetic electron with an atom in the ground state. The impact of the electron causes it to exchange some of its energy with the atom and the latter becomes excited. This process can be represented by means of the following equation:

$$A + e = A^* + e, \tag{3.1}$$

where $A$ represents the atom in its ground or lower state, $e$ the electron and $A^*$ the atom in its excited state.

A collision of the second kind involves the collision of an excited atom in a metastable state with another atom of a different element in an unexcited state, the final result being a transfer of energy from the metastable state to the excited state. The metastable atom therefore loses energy and reverts to a lower state. A collision of the second kind is represented by the following equation:

$$A_1 + A_2^* = A_1^* + A_2, \tag{3.2}$$

where $A_1$ is the unexcited atom of an element labelled 1 and $A_2$ is an unexcited atom of an element labelled 2, the asterisks indicating excited states.

### (*a*) *The helium–neon laser*

The first gas laser to be constructed was a helium-neon laser built by Javan in 1960. It depends for its operation on a population inversion between two excited levels of the atom neon. An energy level diagram of the helium–neon system is shown in fig. 3.4.

The neon bands shown actually consist of many lines and so give rise to many wavelengths of which only the three most powerful ones are indicated.

181

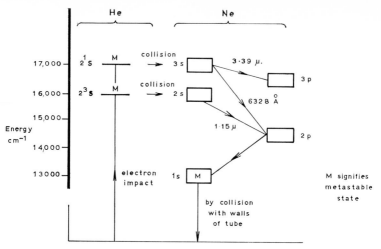

Fig. 3.4. Helium–neon laser energy level diagram.

The helium atoms in the mixture of helium and neon are excited to the metastable states $2^1S$ and $2^3S$ by collisions of the first kind caused by an electrical discharge applied to the laser tube which is normally some 30 to 50 cm long. This process can be represented as follows:

$$He + e = He^* + e. \tag{3.3}$$

These two energy levels are very close to the $2s$ and $3s$ levels of neon and collisions of the second kind then take place leaving the $2s$ and $3s$ levels well populated. This is represented as follows:

$$He^* + Ne = He + Ne^*. \tag{3.4}$$

Javan's first laser used the transitions $2s$ to $2p$ for laser action and a variety of wavelengths were obtained in the near infra-red, the strongest line being at $1\cdot1523$ $\mu$m. Soon afterwards the familiar red line at 6328 Å which arises from a transition between the $3s$ and $2p$ states was discovered. It was also observed, at about the same time, that a very strong output could be obtained further out in the infra-red at $3\cdot39$ $\mu$m the transition $3s$ to $3p$ being responsible. All these transitions involve energy levels which are well above the ground state and so the helium–neon laser is a four-level laser.

It can be seen that the $3\cdot39$ $\mu$m transition shares a common upper level with the 6328 Å transition and so the $3\cdot39$ $\mu$m line will have an adverse effect on the power available in the visible line. Because of its much longer wavelength the threshold of the $3\cdot39$ $\mu$m line is much lower and hence its gain is always much higher compared with the 6328 Å line. It is so high, in fact, that laser action can be obtained without the feedback provided by the laser mirrors being necessary, this effect being called superradiance.

Atoms in the terminal level $2p$ decay radiatively to the $1s$ metastable state in $10^{-8}$ s, which is much faster than the spontaneous rate of decay from the $2s$ to the $2p$ level which has a lifetime of about $10^{-7}$ s; the lower lasing level is thus kept comparatively empty and so the conditions for population inversion with respect to the $2s$ level are satisfied. It is inevitable that some population of the upper levels of neon will take place by direct excitation through collisions of the first kind. Thus the preferential filling of $3s$ and $2s$ levels which occurs by virtue of collisions of the second kind with the metastable helium atoms will be suppressed. Because of this it is advantageous to increase the latter effect in comparison with the former by increasing the density of the helium atoms with respect to the neon atoms. This is achieved, in practice, by filling the laser tube with a pressure of 1 torr of helium and 0·1 torr of neon. An additional important factor to consider is that the $1s$ state of neon is also metastable and so its population will tend to increase until, if it becomes too high, photons emitted by decay from the $2p$ level to the $1s$ level will have a high probability of exciting atoms in the $1s$ level back up to the $2p$ level again. This process is called radiation trapping and is similar to resonant radiation mentioned in Chapter 1. In this case the lifetime of the $2p$ levels is effectively increased, to the detriment of the efficiency of laser action, since a reduced population inversion will result. The population of the metastable $1s$ state therefore has to be reduced by another process which, in order to avoid trapping between the ground and $1s$ levels, must be non-radiative. This is achieved by collision with the walls of the laser tube. It is for this reason that the gain of the helium–neon laser has been found to be inversely proportional to the diameter of the laser tube, so that small tubes of a few millimetres or so in diameter are used. A diagram of a typical helium-neon laser is shown in fig. 2.6. The Doppler width of the 6328 Å transition is 1700 MHz at room temperature so for ordinary tube lengths of tens of centmetres several axial modes will oscillate.

Excitation may be either by means of a radio frequency or a d.c. discharge. For an r.f. discharge external elecrodes are used and power at 27 MHz is supplied, as readily available oscillators form convenient supplies. However, r.f. discharges are liable to cause interference with other radio communications and also the strong fields round the electrodes tend to drive the gases into the walls of the tube so that the pressure drops and eventually the tube must be refilled. Most commercial helium–neon lasers employ a d.c. discharge of 5–50 mA from internal electrodes. The mirrors forming the resonating cavity of a helium–neon laser will be typically a 23 layer dielectric mirror of reflectivity 99·9% and a 9 layer dielectric mirror of transmission 1% which forms the output mirror—assuming, of course, that the laser is to work at 6328 Å which is usually the case.

In comparison with most other lasers the helium–neon laser provides a cheap source of high quality laser light but suffers from the drawback of small power output. The practical upper limit of available power is roughly 100 mW for each of the three major lines. A typical small helium–neon laser gives a uniphase output of about 5 mW. Higher powers are obtainable by pulsed operation; 100 W for a few microseconds have been obtained in this way.

### (b) The argon laser

Unlike the helium–neon laser the argon laser works on a transition between two energy levels of the ionized atom. In order to singly ionize argon atoms, i.e. to remove one electron from each atom, a considerable amount of energy must be supplied to the argon gas. In consequence the power supplies for an argon laser are bulkier and more complex than for the helium–neon laser although the power outputs available are very much higher. However the overall efficiency of 0·05% is about the same.

Figure 3.5 shows the energy level diagram of the argon ion laser and indicates two of the wavelengths in the visible which are available.

A much simplified idea of the mechanism by which atoms are brought into the appropriately excited state can be summarized by the following equation representing a collision of the first kind:

$$e + Ar = (Ar^+)^* + 2e. \qquad (3.5)$$

Depletion of the lower laser level is brought about by radiative decay to the ground state of the ionized atom followed by recombination with an electron to form the neutral atom.

Of the various lines which are emitted, about 80% of the total output power is approximately equally divided between the 4880 Å and 5145 Å lines. The Doppler widths of the argon lines are about 3500 MHz wide, so for a given cavity length more axial modes will oscillate in comparison with a helium–neon laser.

Figure 3.6 shows a diagram of a typical argon laser. The discharge tube, lade of silica, is 60 cm long and has a bore of 3 mm diameter. It is filled with argon gas at a pressure of 0.2 torr. A stabilized current of about 20 A at 300 V is passed down the tube—a current density of some 250 A cm$^{-2}$ being obtained. Such high current densities are necessary in order to ionize sufficient atoms, each of which require an excitation energy of 35·5 eV. A getter is incorporated to remove foreign gases and, owing to ion bombardment of the walls of the tube gas is lost and a reservoir is supplied to compensate for a gradually falling gas pressure within the laser discharge tube. Because the device is d.c. operated positive ions can accumulate at the cathode end of the tube and, being heavier than electrons, the gas pressure tends to increase at that end; a by-pass tube is therefore provided to maintain an even pressure along the tube. A longitudinal magnetic field is applied to the discharge, so

184

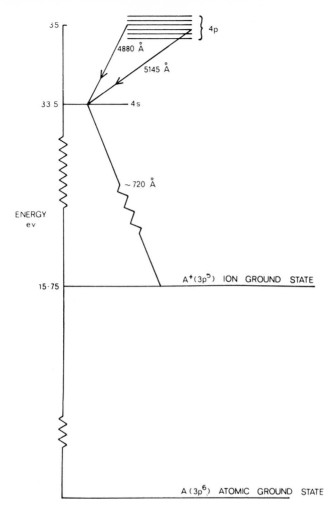

Fig. 3.5. Argon laser energy level diagram.

increasing the electron density for a given current with a consequent enhancement of output power. The laser described has a typical power output for all the wavelengths combined of 1 W, of which approximately 400 mW are available in each of the lines 4880 Å and 5145 Å. Generally speaking, all the lines are produced simultaneously with the usual mirror arrangement, although they all have widely different powers and thresholds, and so a prism cut at Brewster angles is incorporated as part of the cavity; this makes the losses very high for all the wavelengths except one and the laser can be easily tuned to the desired wavelength by simply rotating the end mirror about a vertical axis.

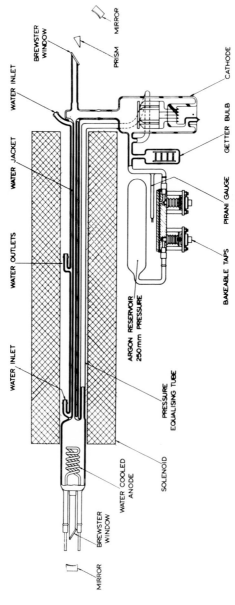

Fig. 3.6. The features of an argon laser (S.E.R.L. diagram).

MIRROR

BREWSTER
WINDOW

WATER COOLED
ANODE

SOLENOID

WATER INLET

WATER OUTLETS

WATER JACKET

WATER INLET

BREWSTER
WINDOW

PRISM

MIRROR

CATHODE

GETTER BULB

PIRANI GAUGE

BAKEABLE TAPS

ARGON RESERVOIR
250mm PRESSURE

PRESSURE
EQUALISING TUBE

Fig. 3.7. A beryllia-tube argon laser (S.E.R.L. photograph).

187

The output power is greatly dependent on current density and considerable increases in power can be obtained by a relatively small increase in current. The latter is, however, limited by the rate at which heat can be dissipated, the anode and discharge tube are therefore usually cooled by the circulation of water. Materials, such as beryllia, which have a higher thermal conductivity than silica are also used in order to pass even greater currents down the tube. Figure 3.7 is a photograph of a beryllia tube laser giving a total power output of several watts.

(c) *The carbon dioxide laser*

This laser is by far the most efficient gas laser and the most powerful continuously operating laser. Efficiencies of more than 20% and outputs of several kilowatts are possible. The carbon dioxide laser works in the infra-red at a wavelength of 10.6 $\mu$m.

In order to appreciate the theory of the carbon dioxide laser it is necessary to discuss the energy levels of the carbon dioxide molecule. It will be apparent that in order to obtain laser action in the infra-red, energy levels whose separation is comparatively small must be found. Suitable levels are found in molecules which do not depend on electron excitation but on the quantization of the vibrational and rotational movements of the molecule. The carbon dioxide laser actually uses two additional gases, nitrogen and helium, the role of nitrogen being similar to the role of helium in the helium–neon laser. The reason for using helium as well will be discussed later.

The carbon dioxide molecule can be pictured as three atoms which usually lie on a straight line, the outer atoms being of oxygen with a carbon atom in between. There are three possible modes of vibration, in each case the centre of gravity remains fixed:

(1) The oxygen atoms may oscillate at right angles to the straight line—this is called, for obvious reasons, the bending mode.

(2) Each oxygen atom can vibrate in opposition to the other along the straight line. This mode is called the symmetric mode.

(3) The two oxygen atoms may vibrate about the central carbon atom in such a way that they are each always moving in the same direction—this is known as the asymmetric mode.

Figure 3.8 shows diagrammatically the three modes, the carbon atom being represented in black.

Each possible quantum state is labelled as follows: For the symmetric mode by 100, 200, 300, etc.; for the bending mode by 010, 020, 030, etc., and for the asymmetric mode by 001, 002, 003, etc. Combinations of all three modes are possible, e.g. 342, but they need not concern us.

In addition to these vibrational modes the molecules can rotate and therefore quantized rotational energies are possible; a set of rotational levels is associated with each vibrational level, these are labelled in order of increasing energy by so called J values, each value being either 0 or a positive integer.

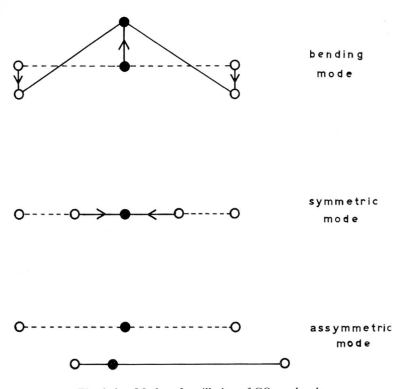

Fig. 3.8.   Modes of oscillation of $CO_2$ molecule.

To make this nomenclature clear fig. 3.9 shows the sets of energy levels associated with each mode of vibration together with a set of rotational levels for the 001 and 100 modes on a much expanded scale. The ground state and the first excited state of the nitrogen molecule are also shown.   As only two atoms are involved the nitrogen molecule can have only one vibrational mode.

The mechanism of laser action is as follows: Direct electronic excitation of the nitrogen molecule into its first excited state by a collision of the first kind.   This process is represented as follows:

$$e + N_2 = N_2^* + e. \qquad (3.6)$$

A collision of the second kind with a carbon dioxide molecule in the ground state with excitation to the 001 state follows:

$$N_2^* + CO_2 = N_2 + CO_2^*(001) \qquad (3.7)$$

This takes place because, as can be seen from the energy level diagram, the two energy values almost coincide.   The 100 vibrational state is of much lower energy and so cannot be populated by this process.

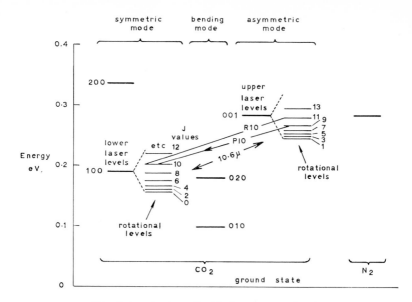

Fig. 3.9. Carbon dioxide laser energy levels.

The population of the 001 levels now exceeds the population of the 100 levels and so the population inversion condition for laser action to take place between these levels has been achieved. However two points must be borne in mind. First, a transition from the 001 level to the 100 level must obey a selection rule which states that $J$ can only change by $\pm 1$. Thus if $J = 10$ for a particular level then only the transitions from $J = 9$ to $J = 10$ and $J = 11$ to $J = 10$ are permitted. If $J$ changes by $+1$ the transition is called a P-branch transition and if $J$ changes by $-1$ it is called an R-branch transition. The line resulting from a transition from $J = 9$ to $J = 10$ is called P10 and that from $J = 11$ to $J = 10$ is called R10. Second, the population of the rotational levels of the 001 state will have a Boltzmann distribution, so, after taking degeneracy into account the effective population of a $J = 11$ level, for instance, will be less than the $J = 9$ level. The result of this is that P-branch transitions dominate because it so happens that a particular P-branch level will fill up (in order to restore equilibrium) by depletion of the population of the R branch above it quicker than the R-branch level population decays by spontaneous emission to the lower laser level. The wavelengths associated with the most powerful transitions of the carbon dioxide laser at normal operating temperatures are: P18—10·57 $\mu$m, P20—10·59 $\mu$m, P22—10·61 $\mu$m and the separation between each transition is about 55 GHz, (55 000 MHz).

Each gain curve corresponding to a P-branch transition has a linewidth of about 50 MHz. This narrow Doppler width, in comparison

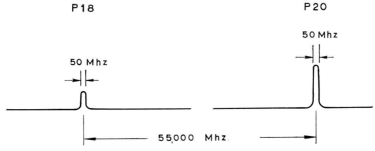

Fig. 3.10. Gain curve of $CO_2$ laser output.

with other gas lasers, comes about because the wavelength is some twenty times as long and the mass of the molecule is greater than that of most atoms. Reference to equation (2.43) will immediately show that these factors will reduce the Doppler width considerably. The sum of the areas under each gain curve in fig. 3.10 is proportional to the population inversion between the 001 and the 100 levels and hence proportional to the intensity of the output. These areas are not in fact equal and it so happens that because of the relative $J$-level populations the area under the P20 gain curve is largest. The axial mode separation for a 100 cm long cavity is by equation (2.62) about 150 MHz. Figure 3.10 shows the P18 and P20 gain curves and the axial mode spacing.

It is apparent from fig. 3.10 that where a tube 1 m in length is used, only one axial mode can oscillate under a gain curve at any given time. If a much longer cavity were to be used the modes would be closer together and so several would oscillate. The axial mode which experiences the greatest gain will tend to grow in intensity at the expense of the others. This happens because the mode which starts to oscillate initially depletes the population of the appropriate 001 level and, as explained above, it so happens that the relaxation rate into such a depleted level from other $J$ levels associated with the same vibrational level (in order to restore a Boltzmann distribution) is much faster than the spontaneous decay rate from any $J$ level to a lower vibrational level. Hence the inversion between other levels tends to feed into the first. The gain profiles will uniformly decrease together and it follows therefore that the P-branch transitions are effectively homogeneously broadened. For a short cavity where only one mode oscillates, the change in cavity length due to instabilities will cause the output power to fluctuate. If the laser is tuned so that the axial mode frequency is at the centre, for example, of the P20 gain curve, then a gradual reduction in power will be observed as the axial mode frequency drifts. If the next mode peaks up at P18 or P22 it will take over, so not only does the power fluctuate, but a frequency fluctuation is also obtained. On the other hand a 10 m cavity with a corresponding mode separation of 15 MHz will result in

191

a number of modes being present under each gain curve, and so the P branch with a maximum gain always oscillates because one axial mode will always be present under the Doppler gain curve.

Inspection of equation (2.43) will show that by operating at a low temperature the Doppler width of the gain curve is kept small and so the gain of the laser line is increased. The temperature is minimized by flowing the gas mixture through the tube so as to change it completely at least twice a second (and in addition removing unwanted carbon monoxide and oxygen resulting from the inevitable decay of some of the carbon dioxide), by restricting tube diameter to a few centimetres and by adding helium to the mixture of carbon dioxide and nitrogen. The helium is effective in (a) increasing the thermal conduction to the walls of the tube, (b) indirectly depleting the population of the lower laser level 100 which is linked through resonant collisions with the 020 and 010 levels, the latter level being directly depleted by the helium and (c) by ' cooling ' the 001 rotational levels which results in the available population being more heavily distributed among the upper lasing levels.

The longer the laser tube the greater the power output—an output of 7 kW has been obtained from a tube 600 ft long which was ' folded ' fifteen times by means of mirrors. Such tube lengths are of course extraordinary and normally about 60–70 W are obtained from each metre of tube. The wavelength of the output brings problems in terms of absorption by laser components. Brewster windows of sodium chloride may be used or the mirrors are attached by means of flexible metal bellows to the tube which is made of glass or silica. The totally reflecting mirror is made by evaporating gold onto stainless steel—the latter acting as a heat-sink which is necessary because of the high power densities. The output mirror is usually of germanium and has a transmission depending on the length of the tube—for a 1 m tube a transmission of about $20\%$ is usual. The germanium is coated on one side with a multi-layer dielectric mirror and on the other with an anti-reflection coating to suppress undesirable reflections between the front and back surfaces. For high power work the absorption of germanium is too high, so mirror substrates made of gallium arsenide are preferred. High voltages of typically 7 kV d.c. with a current of 36 mA are used in order to provide a discharge in a 1 m long tube of 15 mm bore. Typical gas pressures are: 1 torr $CO_2$, 1·5 torr $N_2$ and 5 torr He.

Unlike other gas lasers, the long lifetime of molecules in the upper lasing state means that $Q$-switching is possible and peak power outputs up to $10^4$ times greater than that obtained continuously are obtainable.

Figure 3.11 shows a typical carbon dioxide laser.

### 3. Semiconductor lasers

A brief discussion of the basic principles underlying the conduction of electricity is necessary in order to explain the fundamentals of the semiconductor laser. The action of this laser is quite different from

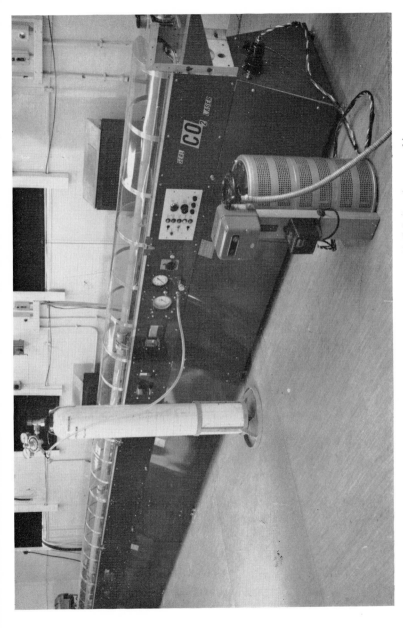

Fig. 3.11. A 450 W carbon dioxide laser (S.E.R.L. photograph).

those which have been considered previously although the fundamental necessity for establishing a population inversion between two energy levels still holds.

We have seen that an atom in free space can be represented by a set of sharp energy levels corresponding to possible energy states of the atom or the possible energy states of the electrons of each atom. When two atoms are in close proximity, each set of energy levels is displaced slightly with respect to the other; each level is said to have been split. If a very large number of atoms are in close proximity, as is the case in a solid such as a crystal, then in general, the energy levels associated with one free atom will split up into as many levels as there are atoms in the body. Figure 3.12 shows the situation for two energy levels in the free atom. Each energy level becomes a band of levels in the solid state.

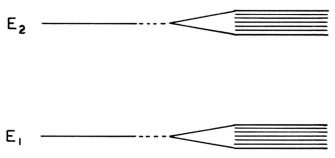

Fig. 3.12. The splitting of two atomic energy levels into bands in a crystal.

Conduction takes place by the movement of electrons under the action of an applied electric field. The field imparts energy to the electron which jumps from one allowed energy level within the band to another higher up. This, in turn, results in a vacant space or 'hole' being left behind into which another electron from lower down can jump. This process continues and constitutes the passage of an electric current. For most solids, a large number of bands will exist with the electrons occupying bands having the lowest energy first. Eventually at a sufficiently high energy a band will exist which is either partially filled or completely filled. Such a band is called the valence band. In each case the next higher band will be empty and is called the conduction band. At room temperatures, of course, this is not strictly true because thermal energy will result in some electrons occupying the conduction band.

In the situation where the valence band is only partially filled, electrons can easily move up to the adjacent higher energy levels within the band and so support conduction. This is the situation with metals such as copper which are good conductors. If, however, the valence band is completely full, electrons will not be available for conduction unless they can reach the empty conduction band. Every material conducts to some extent because, as Boltzmann's equation shows, there is always a finite probability of thermal excitation to any higher level, but as the thermal energy at room temperatures $kT$ is only about 0·025 eV, the energy gap between the conduction and the valence band must be reasonably small. A class of materials called intrinsic semiconductors exists in which the separation between the conduction and valence band is about 1 eV. This results in some electrical conductivity being possible as a reasonable number of electrons are thermally excited to the conduction band (the number can be calculated by the Boltzmann distribution equation given earlier). Materials other than conductors and semiconductors have valence bands separated from conduction bands by considerably larger energy differences—very few electrons are in the conduction band and hence these materials are insulators. The conductivity of semiconductors falls somewhere midway between that of conductors and insulators. This band theory of conduction accounts most satisfactorily for the observation that the conductivity of semiconductors increases with temperature (due to more electrons being thermally excited from the valence band into the conduction band) while the conductivity of ordinary conductors falls with increasing temperature (because increasing the temperature does not increase the number of electrons available for conduction, it only increases their probability of collision and hence the resistivity).

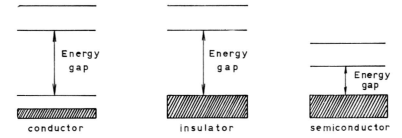

Fig. 3.13. Valence and conduction bands for three different types of materials. Shaded areas indicate levels occupied by electrons.

Figure 3.13 shows each of the cases described above. In order to avoid confusion, no electrons are shown in the conduction band of the semiconductor—this will only happen in reality at the absolute zero of temperature when a semiconductor will obviously be an insulator.

The semiconductor laser makes use of what is known as an extrinsic semiconductor. If an intrinsic semiconductor has atoms of some foreign material diffused into it by a process called doping, its energy diagram acquires new levels which lie between the conduction and valence bands. Semiconductors with these additional levels are said to

195

be extrinsic. When a material, whose atoms have one valence electron more than the atoms forming the host lattice, is diffused into the host lattice, the result is called an n-type material. Spare electrons will therefore become available and will give rise to energy levels close to, but just below, the conduction band (in the case of silicon doped with phosphorus the energy gap is about 0·05 eV). Such an impurity is said to result in donor levels as it enables the material to act as a source of electrons available for conduction.

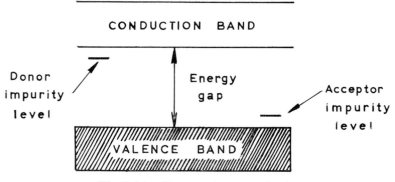

Fig. 3.14. Donor and acceptor levels.

Figure 3.14 shows one such donor impurity level. If there are $10^{22}$ atoms per cubic centimetre and 0·1% doping is used then there will be $10^{19}$ such donor levels per cubic centimetre. The effects of such levels are considerable because thermal excitation will easily populate the conduction band from the donor levels enabling conduction to take place by means of transfer between energy levels in the conduction band. In a similar manner the host can be doped with a material whose atoms have one electron less than those of the host atom. This type of doping produces what are called p-type materials. With a p-type material positive holes are created whose energy levels lie close to, but just above, the valence band and are known as acceptor levels (in the case of silicon doped with aluminium the energy gap is about 0·08 eV). Electrons are then easily thermally excited from the valence band into the impurity levels and conduction can take place by electron movement within the conduction band (see figure 3.14).

A semiconductor laser is made by forming a junction between p- and n-type materials in the same host lattice so as to form what is known as a p-n junction. The doping is extremely heavy, of the magnitude indicated above, so that the lower part of the conduction band of the n-type material is actually filled with electrons and the top part of the valence band of the p-type material is filled with holes. A voltage sufficiently high to overcome the energy gap V shown in fig. 3.15 is then applied to the junction so that the n-type region is connected to a negative

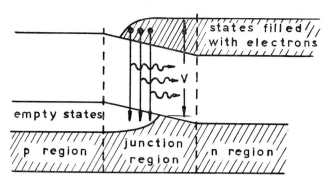

Fig. 3.15.   Semiconductor laser–junction region.

supply and the p-type region to a positive supply.   Under these conditions the junction is said to be forward biased.   The electrons in n-type region and the holes in the p-type region are then driven towards the junction where they recombine to produce photons.

Without the forward bias few photons would be produced as an electron would have to climb a potential barrier before it could recombine with a hole to form a photon.   As the forward bias is increased, more and more photons are emitted and so the light intensity becomes stronger.

The production of photons in this way was known before the advent of the semiconductor laser and is an example of electroluminescence.   It is particularly strong in the material gallium arsenide.   In 1962 several groups of workers reported laser action by passing extremely high currents and simply polishing the ends of the gallium arsenide p–n junction so that they acted as laser mirrors.   Such a gallium arsenide laser is shown in fig. 3.16.

Each end of the junction is polished or cleaned and the sides are roughened to prevent unwanted laser action in other directions and hence wasted population inversion.   The region of the p–n junction where laser action takes place is only a micron or so wide and a millmetre across.   As the current across the junction is increased, spontaneous emission over a wide line-width gives way to laser action at some threshold current and the line-width narrows dramatically.   The threshold current density is constant at about 200 A cm$^{-2}$ below 20°K and rises rapidly as the temperature is increased.   At liquid nitrogen temperatures (77°K) the threshold is 750 A cm$^{-2}$, while at room temperatures 30 000 A cm$^{-2}$ is required, so that the device is usually cooled for more efficient operation as well as removal of dissipated heat.

The energy gap for gallium arsenide V is 1·47 eV corresponding to a wavelength of 8400 Å.   For a particular device the gap varies in width by about 20 Å so that although the very close mirrors result in an axial mode

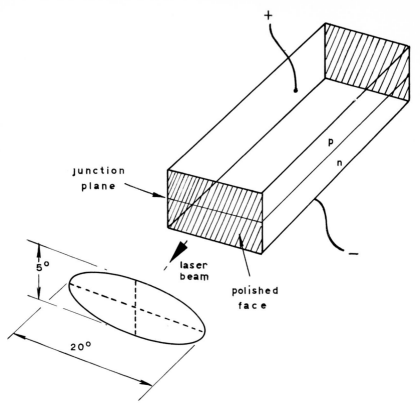

Fig. 3.16. The semiconductor laser.

spacing as wide as 3 Å for a 1 mm mirror separation, many axial modes oscillate. Line-widths of individual modes of less than 300 MHz have been obtained. The losses for transverse modes are very small and the output beam has considerable divergence as indicated in fig. 3.16.

Two efficiencies can be quoted for a semiconductor laser. First, the percentage of photons obtained with respect to the number of electrons crossing the junction; this is called the external quantum efficiency and varies with temperature. At room temperatures it is 15%, at 77°K 40%, while at liquid helium temperatures it can be as high as 60%, and second, in terms of the percentage of laser power output with respect to electrical power input the efficiency can be as high as 10%.

Continuous power outputs of 1 W have been obtained at 20°K, while at 77°K only milliwatts are available continuously although pulsing at this temperature has produced 100 W peak. Only very low powers have been obtained continuously at room temperatures and pulsing is normally necessary because overheating is always a problem. 20 W of peak power have, however, been produced at room temperatures.

Because the stimulated emission arises from within the host material and not from a small percentage of the active medium as in other lasers, the semiconductor laser is by far the most powerful for its size. A wide variety of materials with different dopings can be used and so a wide range of wavelengths is available extending from 3500 Å to 20 $\mu$m. Semiconductor lasers can be tuned by varying an applied magnetic field, the temperature or the pressure and phosphorus-doped gallium arsenide lasers can be tuned by varying the degree of doping.

# INDEX

## Introduction

## Part I

# Part II

# Part III

206

# THE WYKEHAM SCIENCE SERIES
*for schools and universities*

Price per book for the Science Series **20s—£1.00 net.** *in U.K. only*

# THE WYKEHAM TECHNOLOGICAL SERIES
*for universities and institutes of technology*

Price per book for the Technological Series **25s.—£1.25 net** *in U.K. only*
*Standard Book Catalogue numbers.